AF352654

SIGNS OF THE MATERIAL WORLD

Signs of the Material World

*Dostoevsky, Science, and the
Nineteenth-Century Novel*

MELISSA FRAZIER

UNIVERSITY OF TORONTO PRESS
Toronto Buffalo London

© University of Toronto Press 2025
Toronto Buffalo London
utorontopress.com
Printed in the USA

ISBN 978-1-4875-6070-6 (cloth)   ISBN 978-1-4875-6072-0 (EPUB)
                                 ISBN 978-1-4875-6071-3 (PDF)

Library and Archives Canada Cataloguing in Publication

Title: Signs of the material world : Dostoevsky, science, and the
   nineteenth-century novel / Melissa Frazier.
Names: Frazier, Melissa, 1965– author
Description: Includes bibliographical references and index.
Identifiers: Canadiana (print) 2024044292X | Canadiana (ebook)
   20240442962 | ISBN 9781487560706 (hardcover) | ISBN 9781487560720
   (EPUB) | ISBN 9781487560713 (PDF)
Subjects: LCSH: Dostoyevsky, Fyodor, 1821–1881—Criticism and
   interpretation. | LCSH: Monism in literature. | LCSH: Science in
   literature. | LCSH: Russian fiction—19th century—History and
   criticism. | LCSH: English fiction—19th century—History and
   criticism. | LCSH: American fiction—19th century—History and criticism.
Classification: LCC PG3328.Z6 F73 2025 | DDC 891.73/3—dc23

Cover design: Val Cooke
Cover image: iStock.com/Trifonov_Evgeniy; The Print Collector/Alamy
Stock Photo

We wish to acknowledge the land on which the University of Toronto
Press operates. This land is the traditional territory of the Wendat, the
Anishnaabeg, the Haudenosaunee, the Métis, and the Mississaugas of the
Credit First Nation.

University of Toronto Press acknowledges the financial support of the
Government of Canada, the Canada Council for the Arts, and the Ontario
Arts Council, an agency of the Government of Ontario, for its publishing
activities.

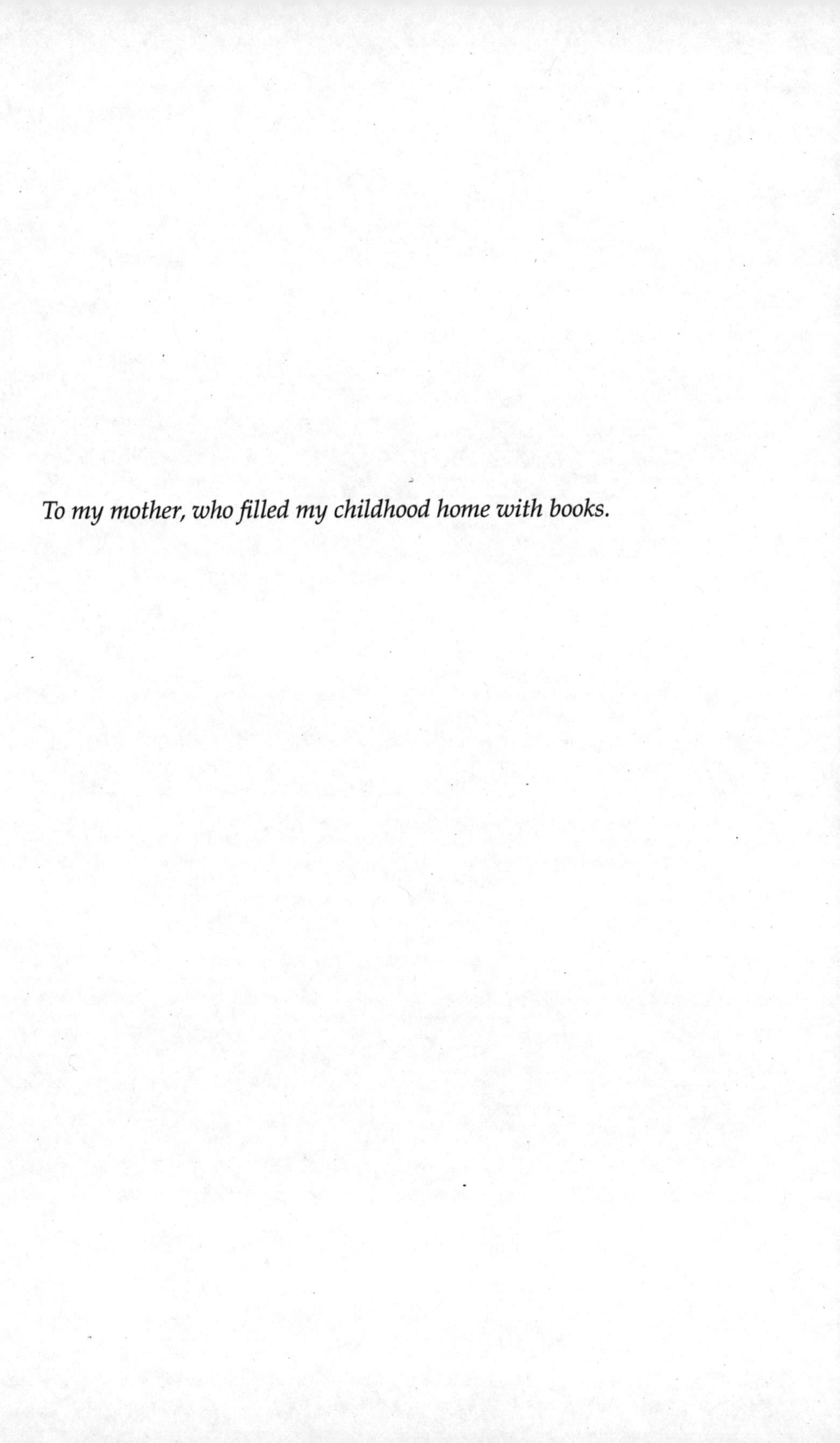

To my mother, who filled my childhood home with books.

Contents

Acknowledgments

I am fortunate to teach at an institution that privileges undergraduate education, and *Signs of the Material World* owes an enormous debt to my brilliant, brave, and independently minded Sarah Lawrence students. For many years, even as my students clamoured for more Dostoevsky, I resisted the idea of a Russian literature defined by Big Ideas. It was only when I finally gave in and taught a course on "Dostoevsky and the 1860s" that I realized that Dostoevsky is not just a writer of astonishing formal inventiveness, but also often wildly funny. When I promptly followed up with another on "Dostoevsky and the West," I discovered in Dostoevsky an extraordinary reader. It was our discussion of Nabokov's *Lolita* (1955) in the latter that started me thinking about "Science, with a capital S," and I am especially grateful to the students who joined my next Dostoevsky class, "Dostoevsky in an Age of Positivism." Even as this book project comes to an end, I am still unfolding the implications of their rigorous and engaged thought.

I am also grateful to the many colleagues who supported me through this long process of learning. Well before 19v, Ilya Kliger and Anne Lounsbery invited me to present on Dostoevsky and Wilkie Collins at the NYU Russian and Slavic Studies Colloquium, and I still remember the free-wheeling conversation that followed, as well as the graduate student in the history of science who first introduced me to the work of John Tresch. I also owe a very special debt of gratitude to the late Deborah Martinsen. It was Deborah who first found a place for my work on "sensation" in *Dostoevsky Studies* and who then attended my many talks over the years, including a presentation on Dostoevsky and Hawthorne at the 2019 International Dostoevsky Society Symposium in Boston; I can't imagine this project without her warm welcome and generous support. I also give a deep and heartfelt thanks to Katia Bowers and Kate Holland, editors of *Dostoevsky at 200: The Novel in Modernity* (University of Toronto Press, 2021). While I admire all aspects of Katia and Kate's work, I thank them above all for their commitment to collaboration, and it is through their good offices that I met many scholars of Dostoevsky whose innovative work continues to inform my own, including Greta Matzner-Gore, Chloë Kitzinger, and Vadim Shneyder.

Among my Sarah Lawrence colleagues, I want to acknowledge Philip Ording for kindly allowing me to sit in on his class on non-Euclidean geometry, and both Sarah Hamill and Scott Shushan for their invaluable help with Peirce. Thanks, too, to the colleagues who so generously welcomed me into the Sarah Lawrence Interdisciplinary Cluster on the Environment, including Sarah again as well as Heather Cleary and Eric Leveau, and I note that the most recent iteration of my class on Dostoevsky and science, now called, after the book, "Signs of the Material World," was taught under the auspices of the grant that they received from the Mellon Foundation in 2021 as part of the *Humanities for All Times* initiative. Last but far from least, I send a very special thank you to my dear friend, professor of physics and Provost and Dean of Faculty Kanwal Singh. Kanwal was never too busy to lend an ear to my forays into her own field, and this book for me also serves as a reminder of the five years that we spent working together in two offices separated only by the "baby's bathroom."

As I put together an abstract in the waning days of the pandemic, my former student Rohan Kamicheril kindly lent his editorial expertise and his much-appreciated enthusiasm for the project, as did another excellent writer, professor of history and Associate Dean of the College Lyde Sizer. Once at the University of Toronto Press, I found myself in very good hands, and I am especially grateful to my editor, Stephen Shapiro, for his unerring ability to identify exactly that part of my writing that needed more work and for his consummate professionalism. It was also Stephen who lined up three extraordinary outside readers who responded to my first draft with a clear understanding of my aims even as they held my work to a higher standard; I thank all three for their detailed, sharp, and inspiring readings.

On the home front, my husband and now grown-up children have lived with this project as long as I have, and their enduring support means all the world to me. While Lucy loves plot as much as I do, Rose's interests take a more scientific turn, and I read both Richard Prum's *The Evolution of Beauty* (Anchor, 2018) and David Graeber and David Wengrow's *The Dawn of Everything* (MacMillan, 2021) at her recommendation. Joe is the one who turned me on to *The Metaphysical Club* (Farrar, Straus and Giroux, 2002). "I think this has something to do with your stuff," he said. He was right.

Parts of chapters one and two were previously published in Russian as "Nauka realizma" ("The Science of Realism") in *Russkii realizm XIX veka*. Ed. M. Vaisman, A. Vdovin, I. Kliger, and K. Ospovat (Moscow: NLO, 2020), 408–30. Chapters two and three draw on work that first appeared in "The Science of Sensation: Dostoevsky, Wilkie Collins and the Detective Novel," *Dostoevsky Studies, New Series* 19 (2015): 7–28. Chapter four includes work previously published in two different articles, "Minds and Bodies in the World: Dostoevskii, George Eliot, and George Henry Lewes," in *Forum for Modern Language Studies* 55, no. 2 (April 2019): 152–70, and "Allegories of the Material World: Dostoevsky and Nineteenth-Century Science," in *Dostoevsky at 200: The Novel in Modernity*, ed. Katherine Bowers and Kate Holland (Toronto: University of Toronto Press, 2021), 81–98. A brief outline of the thoughts on Dostoevsky and eco-criticism that guide my argument throughout appeared as "Minds and Bodies in the World, or: Learning to Love Dostoevsky," *NYU Jordan Center*, 19v: An Occasional Series on the 19th Century, 10 December 2020, https://jordanrussiacenter.org/blog/minds-and-bodies -in-the-world-or-learning-to-love-dostoevsky. Accessed 19 August 2024. Finally, for an extension of my argument here into theory of comedy, please see my contribution to *Funny Dostoevsky: New Perspectives on the Dostoevskian Light Side*, eds. Lynn Ellen Patyk and Irina Erman (New York: Bloomsbury Academic, 2024), "Sensations of Laughter: Mind and Matter in *The Brothers Karamazov*."

Note on Citation

This book follows the Library of Congress standard transliteration system for Cyrillic, with a few exceptions. Spellings of some names have been adjusted to reflect more familiar North American English spellings (e.g., Dostoevsky instead of Dostoevskii, Grigoriev instead of Grigor'ev). Some proper names have also been adapted to aid in pronunciation (e.g., Alyosha instead of Alesha, Nastasya instead of Nastas'ia). Transliterated Russian in the bibliographic references that appear in parentheticals, notes, or the Works Cited adheres strictly to the Library of Congress system.

All translations are my own unless indicated otherwise.

Not just Feuerbach, but many of the writers quoted here make frequent use of italics in expressing often strongly held views. At no point in the text have I added to or otherwise changed their own use of italics: their emphasis is their own.

SIGNS OF THE MATERIAL WORLD

Dostoevsky, Science, and the Nineteenth-Century Novel

Dostoevsky is not particularly known for his fondness for science of any sort. Indeed, what usually strikes readers instead is an apparently militant hostility. The unnamed narrator in *Notes from Underground* (*Записки из подполья*, 1864), for example, reads Darwin in the most dismissive of terms:

> Once it's proved to you, for example, that you descended from an ape, there's no use making a wry face, just take it for what it is. Once it's proved to you that, essentially speaking, one little drop of your own fat should be dearer to you than a hundred thousand of your fellow men, and that in this result all so-called virtues and obligations and other ravings and prejudices will finally be resolved, go ahead and accept it, there's nothing to be done, because two times two is – mathematics. Try objecting to that. (*Notes* 13; *PSS* 5: 105)

Of course, the Underground Man objects exactly to that kind of mathematics, while Dmitri Karamazov in *The Brothers Karamazov* (*Братья Карамазовы*, 1881) takes aim at chemistry: "And yet, I'm sorry for God!" is his anguished cry, "Chemistry, brother, chemistry! Move over a little, Your Reverence, chemistry's coming!" (*Brothers* 589; *PSS* 15:28). In *Demons* (*Бесы*, 1872) Shatov rails against what he calls "half-science." "Half-science is a despot such as has never been seen before," he tells Stavrogin:

> A despot with its own priests and slaves, a despot before whom everything has bowed down with a love and superstition unthinkable till now, before whom even science itself trembles and whom it shamefully caters to. These are all your own words, Stavrogin, all except the words about half-science; those are mine, because I myself am only half-science, and therefore I especially hate it (*Demons* 251; *PSS* 10: 199).

Diane Oenning Thompson argues that it's not that Dostoevsky, the son of a doctor after all, and himself trained as a military engineer, entirely denied the "benefits of science." Instead, she writes, Dostoevsky "was more concerned to emphasise the dangers," many of which, she maintains, have been realized in our own day: "the 'therapeutic cloning' of human embryos … the human genome project … robotics, cryonics, genetic engineering, 'virtual reality' people, designer babies, artificial intelligence, and on the social level, hedonism, the breakdown of community and a sense of pervasive anonymity" (Thompson 210). More recent studies qualify Thompson's claim, as scholars increasingly heed Anna Schur's call for "a more refined reading of Dostoevsky's view of contemporary science," one that would "restore to the science of Dostoevsky's times some of its intellectual range and complexity" (Kaladiouk 419–20). It is certainly true that Dostoevsky emphasized the dangers of a particular kind of science that we still know well today. Dostoevsky also, however, fully embraced and even advocated a different kind of entirely mainstream nineteenth-century science, one that would return real minds and bodies to the world.

Despite its claims to "objectivity" and to a degree of professionalization evident in the 1833 invention in English of the very term "scientist," the nineteenth century was in fact particularly abundant in different, more flexible ways of approaching the material world. In *The Romantic Machine: Utopian Science and Technology after Napoleon* (2010), John Tresch makes the case for a Schellingian *Naturphilosophie* that never really went away. Tresch begins with a "dominant image of modern science" that depends on what he calls "certain exemplary 'classical machines': balances, levers, and clocks." In contrast, in Paris in the 1820s, 1830s, and 1840s, he writes, "a new set of machines came on the scene: steam engines, batteries, sensitive electrical and atmospheric instruments, improved presses, and photography" (Tresch xi). These "romantic" machines, Tresch argues, brought with them "a new understanding of nature, as growing, complexly interdependent, and modifiable, and of knowledge, as an active, transformative intervention in which human thoughts, feelings, and intentions – in short, human consciousness – played an inevitable role in establishing truth" (xi). While for Tresch, after 1850 "the classical image of science again took the upper hand" (xii), in Gillian Beer's seminal reading of *On the Origin of Species* (1859) in *Darwin's Plots: Evolutionary Narrative in Darwin, George Eliot and Nineteenth-Century Fiction* (1983), something of Tresch's "alternative scientific tradition" continues in Charles Darwin (1809–82). As Beer writes, "Darwin was much wounded by Herschel's description of his theory as 'the law of higgledy-piggledy', but the phrase exactly

expresses the dismay many Victorians felt at the apparently random –
and so, according to their lights, trivialized – energy that Darwin perceived in the natural world." The problem, she explains, is that "Darwinian theory will not resolve to a single significance nor yield a single
pattern. It is essentially multivalent. It renounces a Descartian clarity, or
univocality" (Beer, *Darwin's Plots* 6–7). The rise of a science of physiology also offered more than one way to understand the relationship of
mind to body and mind to material world.

Scholars like Rick Rylance increasingly argue that the British George
Henry Lewes (1817–78) in his partly posthumously published *Problems
of Life and Mind* (1876; 1877; 1879) exactly anticipated the more sophis-
ticated approaches to cognitive science that we associate with our own
day, and Lewes with his wide reading worked under the influence of
other scientists, including German physicist and physiologist Hermann
von Helmholtz (1821–94), famous above all for his work on sound and
his emphasis, not just on the source from which the sound emanates but
also on the receiving capacity of the human ear. While newly emerging
sciences of the human body are particularly striking in their emphasis
on perception as a kind of interaction, this other way of thought also
had implications for the "hard" sciences and even for mathematics. Both
Lewes and Helmholtz contributed to the heated debate over the development of non-Euclidean geometry from the 1860s on, for example, and
their (qualified) dissemination of a mathematics that cuts off from the
world as we know it to imagine other possible kinds of spaces reflects
the same belief that perception and what we might call the material
world mutually inflect one another. As Beer argues, these other kinds of
relationships with the material world were also richly reflected in and
shaped by nineteenth-century literature.

Beer begins with Darwin's own reading, making the case for Darwin's debt as a writer to Milton and also to Charles Dickens (1812–70),
especially in *Bleak House* (1853) "with its apparently unruly superfluity
of material gradually and retrospectively revealing itself as order, its
superfecundity of instance serving as an argument which can reveal
itself only through instance and relations" (Beer, *Darwin's Plots* 6). She
also shows George Eliot (1819–80) as a reader of Darwin in turn. In
Eliot's own day Henry James famously sniffed that *Middlemarch* (1871)
was "too often an echo of Messrs. Darwin and Huxley" (H. James 428).
Beer doesn't disagree, but her Darwin is also not James's. Beer instead
reads the much-quoted passage in *Middlemarch* that begins with a tele-
scope "sweeping" the parishes of Tipton and Freshitt and ends with a
microscope "directed on a water-drop" (*Middlemarch* 59–60) as evidence
of Eliot's belief, like Darwin's, in "the plurality of worlds, scales, and

existences beyond the reach of our particular sense organisation." For Eliot, Beer argues, these different lenses serve as "permitting factors" in a "particular strain of Romantic materialism – a sense of the clustering mystery of a material universe" (Beer, *Darwin's Plots* 142). The same is also true for Dostoevsky.

Like Eliot and Wilkie Collins (1824–89) and even his beloved Friedrich Schiller (1759–1805), Dostoevsky tells stories of minds and bodies functioning in the world as an integral part of what Razumikhin in *Crime and Punishment* (Преступление и наказание, 1866) calls "the *living* process of life" (*Crime* 256; *PSS* 6: 197). Dostoevsky's sense of the human mind and body as aspects of a natural world that enjoys its own agency finds an especially beautiful expression in *The Brothers Karamazov* in Ivan's love for the "sticky little leaves that come out in spring" (*Brothers* 230; *PSS* 14: 210), as in Father Zosima's belief that animals, too, have souls. "But can it be that they, too, have Christ?" a young companion asks. "How could it be otherwise," Zosima says, "for the Word is for all, all creation and all creatures, every little leaf is striving towards the Word, sings glory to God, weeps to Christ, unbeknownst to itself, doing so through the mystery of its sinless life" (*Brothers* 295; *PSS* 14: 268). Like Lev Tolstoy (1828–1910), Dostoevsky attempts to apprehend the multiplicity of what he also calls "living life" (*Notes* 129; *PSS* 5: 178) through a more recent, more complicated, and also more flexible kind of mathematics. Like the many Western European and even American writers that he often knew well, Dostoevsky finally doesn't just tell a different kind of science, but enacts it.

When Mikhail Bakhtin defines the dialogicity that in his famous reading characterizes Dostoevsky's art, he does so in part by arguing that science, in contrast to literature, is inherently monologic. Although he acknowledges that scientific activity does require that one deal with another's discourse – "the words of predecessors, the judgment of critics, majority opinion and so forth" – Bakhtin presents the relationship of the scientist to his or her subject as one-way. Because "[t]he entire methodological apparatus of the mathematical and natural sciences is directed toward mastery over mute objects, brute things, that do not reveal themselves in words, that do not comment on themselves," Bakhtin writes, "[a]cquiring knowledge here is not connected with receiving and interpreting words or signs from the object itself under consideration" (Bakhtin 51). Although literary critics may think otherwise, science in fact isn't always practised in those terms. Where Euclidean geometry, for example, depended on our belief, in Douglas Hofstadter's words, that words like "point" and "line" are "necessarily univalent, capable of only one meaning," non-Euclidean geometry as it developed in fits and starts from the eighteenth century on started exactly from the recognition that "the four postulates of absolute geometry simply do not pin down the meanings of the terms 'point' and 'line'" and "that there is room for *different extensions* of the notions" (Hofstadter 222). My claim here is that Dostoevsky's art *is* his science, in his reliance on plot and a particular relationship with readers' bodies as well as their minds that Wilkie Collins's critics derided as "sensational," as in his recourse to an often extravagant figurative language that both represents and embodies the more complicated relationship of subject or mind to material world that another kind of science entails. While this combined literary and scientific practice reflects Dostoevsky's notably transnational and interdisciplinary reading, it also transforms our own. As Eliot would say, Dostoevsky "changes the lights for us" (*Middlemarch* 762): once drawn into his orbit, Eliot herself no longer looks quite the same.

In chapter one, "Nineteenth-century Materialisms," I start with the most immediate target of Dostoevsky's hostility. What Dostoevsky's Shatov disparages as "half-science" is better known as nihilism, a combination of Auguste Comte (1797–1857) and the "vulgar" materialists together with a dash of Claude Bernard (1813–78) that paradoxically lays claim to both an abstract mathematical rationality and a strict empiricism. The fictional nihilist as Ivan Turgenev (1818–83) first invents him in *Fathers and Sons* (*Отцы и дети*, 1862) and as Nikolai Chernyshevsky (1828–89) then improves him in *What Is to Be Done?* (*Что делать?*, 1863) is a quintessential figure of nineteenth-century Russian literature, so much so that we often associate his stance with Russia alone. As Dostoevsky also reminds us, however, this same jumble of sometimes contradictory thought enjoyed equal popularity across Europe, not just in its separate components, but in the quasi-scientific commitment to material monism that Comte's positivism inspired.

Just as we tend to see the nihilist/positivist world view as a uniquely Russian phenomenon, so we often read Dostoevsky's engagement with another sort of science with an eye to what is distinctively Russian in his response.[1] This chapter emphasizes instead what remains the same. "Nineteenth-century Materialisms" adds to our consideration of Tresch's "alternative scientific tradition," yet another British scientist of worldwide renown, physicist James Clerk Maxwell (1831–79), along with a scientist and science-writer whose influence operated on an admittedly much more local level: Dostoevsky's sometime friend and journalistic co-contributor, Nikolai Strakhov (1828–95). This chapter finally grounds Dostoevsky's particular strain of nineteenth-century materialism in his reading of a philosopher once wildly popular but now largely forgotten: Ludwig Feuerbach (1804–72).

Feuerbach has long been dismissed, in Marx W. Wartofsky's words, as "either as a transitional figure (between Hegel and Marx), as a purveyor of aphorisms ('Man is what he eats,' a pun in German that is not even Feuerbach's own), or as a crude and simple materialist" (Wartofsky 1). In the 1840s and 1850s, however, he made for thrilling reading. As Alexander Herzen (1812–70) recalled his first encounter with *The Essence of Christianity* (*Das Wesen des Christentums*, 1841), "After reading the first pages I leapt up with joy. Down with the trappings of masquerade; away with the stammering allegory" (Kelly 198). For Herzen, as for young people across Europe, Feuerbach's enormous appeal lay in his insistence on a "sensuousness," or *Sinnlichkeit*, that would clear away the clutter of abstract thought and put us in direct contact with material reality, including our own. As Feuerbach himself writes in his preface to the second edition of *The Essence of Christianity*, he "unconditionally repudiate[s] *absolute*, immaterial self-sufficing speculation, that speculation which draws its material from within" (Feuerbach, *Essence* xiv). His is a philosophy, Feuerbach explains, that does not "regard the *pen* as the only fit organ for the elevation of truth, but the eye and the ear, the hand and foot," one that "recognises as the true thing, not the thing as it is an object of the abstract reason, but as it is an object of the real, complete man" (xv). As Feuerbach's vehemence here, as elsewhere, would suggest, a commitment to thought only ever as embodied characterizes not just the content of Feuerbach's writing but also his emotional and often unorthodox style. The same is also true for the novelists that I consider here, above all Dostoevsky.

Chapter two, "Of Doctors and Detectives, Chemistry and Mathematics," starts on the level of content. The nineteenth-century novel is remarkable for its often overt claims to scientific authority, not just in the case of Eliot but when Honoré de Balzac (1799–1850) frames *The Human Comedy* (*La Comédie humaine*, 1842) with reference to zoology, for example, or when Émile Zola (1840–1902) claims the mantle of Bernard in his "experimental" novel. It is also striking for the many scientist-heroes that fill its pages. Collins's arch-villain Count Fosco in *The Woman in White* (1859) is an amateur chemist, while Vautrin as we meet him in Balzac's *Père Goriot* (1835) is an instinctive social scientist; Lydgate in Eliot's *Middlemarch*, like Bazarov in *Fathers and Sons* as well as Lopukhov, Kirsanov, and even Vera Pavlova in *What Is to Be Done?*, is a medical doctor. This scientific bent is perhaps most noted in an emerging genre of detective fiction, from Edgar Allen Poe's (1809–49) Dupin stories through Arthur Conan Doyle's (1859–1930) Sherlock Holmes. Especially as put to work in the service of detection, this science is usually read in nihilist/positivist terms, including as an expression of the aspiration for power that even the more strictly scientifically grounded Bernard sees as a corollary of absolute knowledge. A science that strives for what Chernyshevsky calls "extraordinariness" is not the only kind that the nineteenth-century novel has on offer, however.

In *War and Peace* (*Война и мир*, 1865), his extended reference to calculus offers an admittedly increasingly uncomfortable Tolstoy one last means of reconciling the scientific rigour that he craves with a material world that he can't help but see as marked by multiplicity; in Lawrence Frank's reading, as well as Beer's, the Darwinian underpinnings of Dickens's *Bleak House* lend themselves to the same fully scientific form of irresolution. As Thomas Sebeok and Jean Umiker-Sebeok see it, even Sherlock Holmes's detecting opens itself to other readings. It is certainly true that Holmes bears the obvious markers of "extraordinariness," including in his association with yet another medical doctor, his friend Dr. Watson, as well as in his own practice of experimental chemistry. Even when not operating under the dubious influence of his addiction to cocaine, however, Holmes's nonetheless fully scientific approach to detecting relies less on his own claims to the strictly empirical workings of "deduction" than on the kind of inspired guessing that the Sebeoks call, after American philosopher and mathematician Charles Sanders Peirce (1839–1914), "abduction." A less certain way of scientific knowing is finally still more evidently featured in Collins's and Dostoevsky's contributions to the literature of detecting. In the "bold experiment" that solves the mystery in *The Moonstone* (1868) as in Ivan Karamazov's repeated reference to non-Euclidean geometry, science is cast not as a copy of an already existing material world, but as a chain of transformations that "link us to an aligned, transformed, constructed world" (Latour 79). While the content of their novels accordingly already tells a story of mind itself as part of the natural world that it would consider, this point is made still more strikingly when that science is not named by name but embodied as literary form.

Chapter three, "Bodies and Plots," addresses plot not as content but as literary device. Dostoevsky's critics, like Collins's, were often appalled at plots that they saw as so lurid as to bypass readers' rational minds and appeal directly to their bodies. As Pyotr Tkachev (1844–86) put it in his contemporary review of Dostoevsky's *Demons*, "make [the reader's] hair stand on end, entertain him, amuse or frighten him, but just don't make him think or look up from the page" (Tkachev 75). This chapter reads their recourse to what the British called "sensation" as a reflection of George Henry Lewes's pioneering physiological psychology instead. While Lewes's scientific commitment to mind and body as two parts of the same whole, "the two being," in his words, "as the convex and concave surfaces of the same sphere, distinguishable yet identical" (Lewes, *Problems* 103–4), finds its most complete expression in his partly posthumously published *Problems of Life and Mind*, the same principle is also already at work in his earlier artistic endeavour *Ranthorpe* (1847). A science premised on a more complicated materiality finally also accounts for "sensation" as practised by Lewes's closest intellectual interlocutor: his novelist wife, George Eliot.

More recent critics have drawn attention to Eliot's reliance on the tools of "sensation" not just in the unusual example of *The Lifted Veil* (1859), but even in *Middlemarch*. Like Dostoevsky, although with considerably more restraint, Eliot in *Middlemarch* employs typically "sensational" elements of plot, including, in her case, adultery, murder, drug abuse, and fraught issues of inheritance. Like Dostoevsky's far seedier cast of characters, Dorothea and her friends also model the visceral response that Eliot hopes to elicit: just as Raskolnikov is driven to return to the scene of his crime by his desire to experience once more that "spinal chill" (*Crime* 456; *PSS* 6: 346), so Will, Eliot's narrator tells us, "was made of very impressible stuff," especially in the reaction of his nerves that Eliot repeatedly describes as "electric shock" (Eliot, *Middlemarch* 388, 543). Where Tkachev insists on reading its effects in the most "vulgar" of terms, however, "sensation" as Dostoevsky also practises it reflects the Lewesian assumption that we only ever bring our complete selves to the task of knowing, not just our minds but also our bodies, and the two together, yours and mine. "Sensation" in this sense isn't an imperative directed at readers as materially determined automata because it can't be. It is instead an invitation to share in the novelistic enterprise as the living and breathing entities that we are.

Once mind is cast in Lewesian terms as part of the material world that it would consider, science itself becomes the reading of the mutual and reflexive sort of sign that Peirce calls an "index." An "index," as Peirce would have it, is a material trace of the sort that we already encountered in chapter two, the physical evidence that the detective reads in order to reconstruct a larger context. In the case of plot as a literary device, the indexical quality of a sign is constituted not just in its relationship to its original object in the world but also in its effect on the material being of its receiver. In chapter four, "Metaphor and Allegory," I argue that Dostoevsky's signs take on an indexical cast even as he turns to devices of an apparently more literary and less "existential" sort. Dostoevsky's and Eliot's notably bidirectional metaphors in *Crime and Punishment* and *Middlemarch* already operate in indexical terms as the mutual implication of tenor and vehicle works not to elevate one side over the other but to show substance and idea as always interdependent. The literary effects of a science predicated on the interrelationship of mind and material world are still more striking, however, in Dostoevsky's recourse to a kind of allegory that is at the same time insistent and yet radically unstable in its meaning-making.

Allegory as Coleridge would have it, as "an abstraction from objects of the senses" (Fletcher 16 fn. 29), may seem especially distant from the real, material world that science would consider. Certainly, Dostoevsky openly mocks the Chernyshevskyan reliance on an allegory of the "stammering" sort, the kind that tries and fails to reconcile Chernyshevsky's claims to material monism with his socialist utopian ideals. Allegory as Dostoevsky himself practises it, however, offers not a departure from "existential" reality, but its more adequate expression. As Amanda Jo Goldstein writes in *Sweet Science: Romantic Materialism and the New Logics of Life* (2017) with regard to a different, slightly earlier set of writers, once we take facts themselves as other than "plain," "the epistemic status of poetic language *changes*: the tropes, figures, images and metaphors formerly thought to adorn or obscure the simple evidence of sense may instead index fidelity to it" (Goldstein 9). For Dostoevsky in *Demons* as for Dickens in *Bleak House* and Nathaniel Hawthorne (1804–64) in his own takedown of the socialist utopian project, *The Blithedale Romance* (1852), this change in epistemic status is reflected not least in an allegory that posits multiple meanings in operation at the same time and with the same degree of "objective" reality.

As we trace the implications of a shifting but nonetheless "stubborn" materiality, we recuperate the self-conscious literariness of nineteenth-century realism. As I argue in my fifth and final chapter, "Social Physics," we also find ourselves involved in what might seem a particularly muddled political stance. The advocates of what Bruno Latour calls "Science, with a capital S" (Latour 229) in Turgenev's, Chernyshevsky's, and Dostoevsky's novels, nihilists all, are proto-revolutionaries, fictional precursors to a movement that culminated in the 1917 revolution; the "view" that shaped their science, to borrow a term from Dostoevsky's friend Strakhov, was the first glimmerings of the radical restructuring of wealth and power that would be called socialism and then communism. This stark political clarity eludes the advocates of a more flexible and contingent sort of science, in the case of Hawthorne, so much so that what we find looks a good deal like political expediency instead. Like Eliot, however, Dostoevsky was not so much inconsistent in his political views, as deeply idiosyncratic. At the risk of reducing politics to the level of our most immediate interpersonal relationships alone, their idiosyncrasy also reflects a careful reading of Schiller.

Aileen Kelly in her recent work on Herzen, like Tresch in *The Romantic Machine* and Jane Bennett in *The Enchantment of Modern Life* (2001), draws on the fact of Schiller's early training in medicine to read Schiller as philosopher as a New Materialist ally of sorts. Schiller as playwright, on the other hand, is most often read as Martin Malia would have him, as "the poet of pure ego ... of liberty stripped of all its political and social contingencies" (Malia 43), including by the real Russian revolutionaries that Dostoevsky's characters both reflect and anticipate. In "Social Physics" I argue, however, that it is their tragedy that Karl Moor in *The Robbers* (*Die Räuber*, 1781) as well as both Posa and Carlos in *Don Carlos* (1787) turn away from the "somatized and particularized" (Tresch 78) interactions with real people in the world that would temper their finally deadly abstraction. Although the heroes of the novels that I consider here, from Raskolnikov and Ivan Karamazov to Eliot's Dorothea, learn the very same lesson, they do so without dying themselves. In conventionally heteronormative pairs, as brothers, and even, in *The Woman in White*, as a threesome, they find each other instead.

While even professional readers sometimes find fault with Eliot's increasingly conservative leanings, as literary scholars we are rarely comfortable with Dostoevsky's politics in their entirety, especially as expressed on a larger and more systemic level. There is no doubt that Dostoevsky wrote in direct opposition to nihilist politics as well as science. He is also well known not just for his vocal support of Russian orthodoxy, autocracy, and "universality," but for his expression of openly anti-Semitic, anti-Catholic, and anti-Muslim views, all markers of an extreme right-wing political stance, in Dostoevsky's day as in our own. His continuing appeal to readers at the opposite end of the political spectrum, even in the face of the current crisis in Ukraine, nonetheless speaks to his advocacy at the same time of a very different set of values. While this other Dostoevsky is most apparent when he writes in novelistic rather than "prophetic" mode, as Greta Matzner-Gore notes, at his best, as in his famous "Pushkin Speech" (1880), Dostoevsky even as prophet is capable of characterizing his dreams for the future of Russia and of humanity "as beliefs (rather than as certainties)," and as "possibilities (rather than as guarantees)" (Matzner-Gore, *Dostoevsky* 61). Dostoevsky's particular constellation of political views surely reflects not just his reading of Schiller, but also the unusual trajectory of his life, including his own brief involvement in radical politics, his arrest in 1848, and the ten years of prison and exile that followed. It also offers one last expression of his unwavering scientific commitment to the shifting and contingent realities of real individuals in the world.

Even as my last chapter turns to the politics that a certain kind of science brings in its wake, my focus remains on a project of reading. A nihilist/positivist science that strives to make itself one with a world cast as one in turn has little use for reading and none whatsoever for the literary arts. Science of this sort understands its function instead as a laying-bare of objective truth at whatever cost to "living life": as Bazarov puts it in *Fathers and Sons*, "I'll cut the frogs open and look inside to see what's going on; since you and I are just like frogs, except that we walk on two legs, I'll find out what's going on inside us as well" (Turgenev 15–16). A science that starts with the premise of mind as part of the living and changing world that it would consider, however, understands its functioning in fundamentally literary terms, as a reading of the signs of the material world, human and non-human alike, that both represents and also enacts our always-incomplete reality.

While science as an interactive and ongoing act of reading claims an authority of its own, it grounds that authority not in the would-be certainties of a "brain-in-a-vat," but in what Latour calls "the number of *relations* established with the world" (Latour 4). Readers of nineteenth-century Russian literature are perhaps most familiar with reference to the "environment" as a marker of the nihilist/positivist belief in the fixed and one-way workings of material determinism. As Cary Wolfe writes in "Ecological Poetics" (2021), however, "the term 'environment' reminds us that what counts as 'nature' is always a product of the contingent and selective practices deployed in the embodied actions of a living system" (Wolfe 102). As itself an act of reading, "environment" in this sense is finally also a collective act of writing, what Wolfe calls "this wild, crisscrossing dance of an almost unimaginable heterogeneity of living beings, at different scales and at different temporalities, doing their own thing" (103).[2] In the narrower terms not of signs in the world but of signs on the page, the project of reading that I describe finally entails a "crisscrossing" of its own.

The nineteenth-century writers that I consider in *Signs of the Material World* are at the same time avid readers of a combined scientific and literary discourse that stretches across Europe from Russia to the United States. Darwin, as Beer notes, reads Dickens, while Dickens reads Lewes, as do Peirce and William James (1842–1910). Collins reads Dickens and Balzac. Lewes and Eliot read Darwin and one another as well as Feuerbach, Helmholtz, Schiller, and even, I suspect, the "sensationalist" writing of Mary Elizabeth Braddon (1835–1915). Dostoevsky is the wildest dancer of all, both as an extraordinary reader in his own right and as a representative of a nineteenth-century Russian literary tradition that was itself engaged in a highly self-conscious act of reading.

Dostoevsky with his often parodic and always transformative readings is the motive force behind the literary environment that I delineate here, from Poe and Hawthorne all the way to Chernshevsky, Tolstoy, and Strakhov, as the writer most actively engaged in reading his fellow writers reading the signs of the material world in turn. I am certainly not the first to claim that art offers an ideal medium for both representing and re-enacting the workings of minds and bodies in the world; I am not even first to make a special claim for the nineteenth-century novel in particular.[3] What *Signs of the Material World* offers is a sometimes startling shift in perspective: with Dostoevsky at its unsteady center, our project of reading changes.

1 Nineteenth-Century Materialisms

If Eliot's "sense of the clustering mystery of a material universe" has yet to catch on in the broader scientific imagination, it is not for lack of trying. In the late twentieth and early twenty-first centuries, the movement that calls itself New Materialism offers yet another attempt in that direction. As Jane Bennett writes in an attempt to recuperate a particular relationship with the material world along with a term that Karl Marx seems to have hijacked: "Ilow did Marx's notion of materiality – as economic structures and exchanges that provoke many other events – come to stand for the materialist perspective per se?" (Bennett, *Vibrant* xvi). More often the issue is not Marxist historical materialism, but science itself as a discipline that would understand and explain what we also call the natural world. Where Gilles Deleuze and Félix Guattari's body as "machinic assemblage," like Donna Haraway's "material-semiotic actors," emphasize, in her words, "the object of knowledge as an active part of the apparatus of bodily production" (Haraway 67), for Bruno Latour in works like *Pandora's Hope* (1999) it is a matter of abandoning the absolute certainty of a "brain-in-vat" for a "sturdy relativism, based on the number of *relations* established with the world" (Latour 4), a move that entails granting "nonhumans" their own history and along with it "the multiplicity of interpretations, the flexibility, the complexity" that we have hitherto reserved only for humans (16). As fertile as this field of thought has been in recent decades, however, it is also not entirely of recent invention.

In *Pandora's Hope* Latour is very clear that what he offers is something other than early twentieth-century phenomenology. "For all its claims to overcoming the distance between subject and object – as if this distinction were something that could be overcome! as if it had not been devised so as *not* to be overcome!" Latour writes, "phenomenology leaves us with the most dramatic split in this whole sad story: a world of science left entirely to itself, entirely cold, absolutely inhuman; and a rich lived world of intentional stances entirely limited to humans, absolutely divorced from what things are in and for themselves" (Latour 10). Still, phenomenology, too, explicitly sought a reconciliation of mind and body and mind and world in a new kind of science. As Maurice Merleau-Ponty writes in *Phenomenology of Perception* (*Phénomèlogie de la perception*, 1945):

> All my knowledge of the world, even my scientific knowledge, is gained from my own particular point of view, or from some experience of the world without which the symbols of science would be meaningless. The whole universe of science is built upon the world as directly experienced, and if we want to subject science itself to rigorous scrutiny and arrive at a precise assessment of its meaning and scope, we must begin by reawakening the basic experience of the world, of which science is the second-order expression. (Merleau-Ponty viii)

To achieve that kind of science is to recognize that we are in a relationship with a material world that is also in a relationship with itself. As David Abram explains, for Merleau-Ponty the objects in the world around us "are not merely subjective; they are *intersubjective* phenomena – phenomena experienced by a multiplicity of sensing objects" (Abram 38). If phenomenology takes us back at least to the first half of the twentieth century, Deleuze, like neurologist Antonio Damasio, reaches all the way back to Spinoza and the seventeenth century in an attempt to undo what the latter calls "Descartes's error." In between the two and Marx notwithstanding, the nineteenth century, as my Introduction has already argued, is especially rich in different strains of "new" materialisms, from Tresch's "romantic" machines through non-Euclidean geometry and an emerging science of mind and body. If we often fail to recognize this other kind of science as science at all, it is only in part because, like Bakhtin, we persist in seeing science as the "mastery over mute objects, brute things." It is also because in the nineteenth century Latour's "Science, with a capital S" (Latour 229) spoke in a particularly strident voice: readers of Russian literature know it as Russian nihilism.

Despite the name, Russian nihilism is not "nihilist" in any common understanding of the term, nor even particularly Russian. As advocated most famously in Chernyshevsky's best-selling novel, *What Is to Be Done?*, Russian nihilism instead reflects the same amalgam of popular science and largely left, or at least left-leaning, politics that Western Europe, oddly enough, knows as positivism. By the 1860s Auguste Comte was no longer quite as celebrated in progressive circles as he once was, so much so that Chernyshevsky's heroes briefly adopt "Auguste Comtean" as a term of light abuse. As even their small joke suggests, however, it is above all Comtean positivism that lies behind the nihilist commitment to both mathematical quantification and a strict, if proximate, cause and effect. In *What Is to Be Done?*, Chernyshevsky's narrator also vouches for his hero Kirsanov's fictional credentials with reference to both the real German cell biologist Rudolf Virchow (1821–1902) and the real French physiologist Claude Bernard, the latter especially well-known for his codification in his *Introduction to the Study of Experimental Medicine* (*Introduction à l'étude de la médecine expérimentale*, 1865) of what we would now call the scientific method. The narrator's own claims to the all-powerful effects of material determinism nonetheless derive from the less scientifically grounded but even more popular work of the so-called vulgar materialists, Ludwig Büchner (1824–99), Jakob Moleschott (1822–93), and Karl Vogt (1817–95). In its furthest reaches, the nihilist emphasis on "environment" finally reflects the still more scientifically questionable extension of Comtean thought into the realm of the social sciences, including in the work of Belgian mathematician and statistician Adolphe Quetelet (1796–1874) and British historian Henry Thomas Buckle (1821–62). With the exception of Bernard, Virchow, and to some extent Moleschott, the members of this only very loosely affiliated group practised little in the way of actual empirical research, tending more towards unsupported ideological assertion instead. Even as the scientific establishment then as now gives their claims little credence, however, their craving for scientific certainty persists.

In "Darwin, Dostoevsky, and Russia's Radical Youth" (2016), David Bethea and Victoria Thorstensson update Thompson's list of the contemporary effects of what the latter calls a "heroic materialism" (Thompson 191). "Militant atheism, intelligent design, epigenetics, the selfish gene," they write, "the debate goes on today, on different terms but with equal animation and often vituperation…. We want to believe we are moving forward, but are we not retelling the same narrative with different characters, different facts, different data sets?" (Bethea and Thorstensson 62). As Bethea and Thorstensson conclude: "Enter the Underground Man." As extraordinarily inventive as it is, however, Dostoevsky's response to the nihilist/positivist package also comes with a context of its own. As we reconsider the effects of what Ian Duncan calls "the West's seemingly unlimited imperial and industrial expansion" (Duncan 9) in a twenty-first century marked by war and climate crisis, we do well to keep in mind what has already gone before: even in nineteenth-century terms, science of the nihilist/positivist sort was never the only option.

Nihilists and the Natural World

Any definition of the term "nihilism" can only come with a degree of qualification. In his own response to Chernyshevskyan thought, *The Gift* (*Дар*, 1937), Vladimir Nabokov glosses this "strange word (in which 'nihil,' nothing, corresponds as it were to 'material')" (Nabokov 248) with a definition borrowed from the second edition of Vladimir Dahl's *Explanatory Dictionary of the Living Great Russian Language* (*Толковый словарь живого великорусского языка*, 1863–6; 1880–2; 1903–9): "A scandalous and immoral doctrine rejecting everything that cannot be palpated" (248). Later and more moderate lexicographers tend to argue that the word "nihilist" made its first appearance in Russian literature in Nikolai Nadezhdin's 1829 article "A Throng of Nihilists" ("Сонмище нигилистов") where it seems to refer to a kind of literary barbarism that combines the most obscure aspects of Fichtean thought with a strong dose of Byronic gloom and general excess of emotion. The term only became common after Ivan Turgenev's *Fathers and Sons*, however, and then as a signifier for a different cluster of characteristics.

Fathers and Sons begins with Arkady Kirsanov's return home from the university together with his friend Bazarov. Arkady's father and his uncle are taken aback by Bazarov's unconventional manners, or lack thereof. "Well, and what exactly is this Mr. Bazarov?" Arkady's uncle Pavel Petrovich asks:

– "What is Bazarov?" Arkady grinned. "Would you like me to tell you, Uncle, what he really is?"
– "If you please, Nephew."
– "He's a nihilist."
– "What?" asked Nikolai Petrovich, while Pavel Petrovich raised his knife in the air with a piece of butter on the end of the blade and remained motionless.
– "He's a nihilist," repeated Arkady.
– "Nihilist," said Nikolai Petrovich. "That's from the Latin *nihil*, nothing, as far as I can tell; therefore, the word signifies a person who … acknowledges nothing?"
– "Say, rather, who respects nothing," Pavel Petrovich put in, and once again set about spreading his butter.
– "Who approaches everything from a critical point of view," observed Arkady.
– "Isn't it all the same thing?"

– "No, it isn't all the same thing. A nihilist is a person who doesn't
  bow down before authorities, doesn't accept even one principle on
  faith, no matter how much respect surrounds that principle"
  (Turgenev 18–19).

As Bazarov himself later explains it, whatever negation the word "nihilism" would imply is simply a consequence of the nihilist's essentially
anti-authoritarian, empiricist, and strictly utilitarian bent: "We act on
the basis of what we recognize as useful," he later explains, "Nowadays
the most useful thing of all is negation – we negate" (40, amended). As
both Nabokov and Dahl suggest, however, underlying the young men's
claims to nothing more than an especially strict kind of critical thinking is a particular materialist philosophy. That materialist philosophy is
also the practice of a certain kind of science.

Turgenev later claimed that the real-life model for Bazarov was a pro-
vincial young doctor he met on the Isle of Man, and certainly one of
Bazarov's chief markers is his training in medicine. Bazarov is also char-
acterized by his reluctance to spend much time with actual patients: his
science focuses instead on the dissection of frogs. As Bazarov tells the
peasant boys whom he recruits to help him capture his victims: "I'll cut
the frogs open and look inside to see what's going on; since you and I are
just like frogs, except that we walk on two legs, I'll find out what's going
on inside us as well" (Turgenev 15–16). In Pavel Petrovich's words, "He
doesn't believe in principles, but he believes in frogs" (20). Bazarov
also collects insects and examines "infusoria" (115) through the microscope that he carries in his luggage; he reads and recommends reading,
especially in chemistry, to the various people whom he meets; and he
finally freely dispenses such gems of scientific thinking as "A decent
chemist is twenty times more useful than a poet" (20) or "Nature's not
a temple but a workshop" (35). The same science and scientific outlook
then mark the most famous of the fictional nihilists who followed in
Bazarov's wake, the "new men" of Nikolai Chernyshevsky's *What Is to
Be Done?*

The word "nihilist" never appears in Chernyshevsky's novel. Still,
Chernyshevsky's readers then as now readily recognized the type, not
just in Chernyshevsky's heroes, but in Chernyshevsky himself. Indeed,
for all Turgenev's claims to have drawn Bazarov's features from an
anonymous young doctor, many readers at the time, including Chernyshevsky, saw Bazarov as a kind of caricature of Chernyshevsky himself
or, more likely, of his younger friend and associate Nikolai Dobroliubov (1836–61). Unlike Bazarov, these "real" nihilists were literary

critics with no training in the natural sciences whatsoever. As sons of priests, both Chernyshevsky and Dobroliubov studied at the seminary (Chernyshevsky in Saratov, Dobroliubov in Nizhny Novgorod), before coming to St. Petersburg to study philology (Chernyshevsky in 1846, Dobroliubov in 1853). The education of their fellow nihilist Dmitri Pisarev (1840–68) differed only in that, as a son of the gentry, Pisarev studied first at the gymnasium before graduating in 1861, again from St. Petersburg University and again in philology.[1] All three then made their careers as journalists and literary critics, Chernyshevsky as head of the literary section and then as editor of *The Contemporary* (*Современник*) from 1855 until his arrest in 1862, and Dobroliubov first as contributor to the same journal and then as Chernyshevsky's co-editor until the younger man's untimely death from tuberculosis in 1860. Pisarev wrote mainly first for *The Dawn* (Заря) and then for *The Russian Word* (*Русское слово*) before and *Notes of the Fatherland* (*Отечественные записки*) after his own arrest in 1862 and until his death by drowning in 1868. As the arrests and, in Chernyshevsky's case, another twenty years of internal exile would make clear, for all three, literary criticism served primarily as an expression of radical politics. For all three, and this despite their lack of training, it also served to promote a particular scientific world view.

Pisarev is particularly striking in this regard. Although he wrote widely on literature, including a famous review of *Fathers and Sons* that did a great deal to define Bazarov as a new "type," Pisarev himself, as E. Lampert notes, once described "his main task" as "to acquaint the public with the best in European scientific knowledge" (Lampert 296). Lampert's wonderful term is *"haute vulgarisation,"* as in the long essay on Darwin's *On the Origin of Species*, "Progress in the Animal and Vegetable World" ("Прогресс в мире животных и растений," written in prison in 1864), that Lampert describes as a "first-class work of *haute vulgarisation* which was largely responsible for acquainting the Russian public with Darwinism" (300). While Dobroliubov didn't devote the same attention to issues in science, Chernyshevsky also did his bit for *"haute vulgarisation."* Chernyshevsky devotes a large part of his long review essay "The Anthropological Principle in Philosophy" ("Антропологический принцип в философии," 1860), for example, to presenting chemistry as, "perhaps, the greatest glory of our age" (Chernyshevskii, *Selected* 80) and also a harbinger of things to come:

[T]he union of the exact sciences, under the government of mathematics, that is, counting, weighing and measuring, is year after year spreading to new spheres of knowledge, is growing by the inclusion of newcomers. Chemistry was gradually followed by all the sciences concerned with plant and animal organisms: physiology, comparative anatomy, various branches of botany and zoology. Now the moral sciences are joining them. (91)

The most important nihilist statement on science and scientific method is found not in an article or book review, however, but in Chernyshevsky's famous foray into novel writing.

Chernyshevsky wrote *What Is to Be Done?* as a direct response to what he saw as Turgenev's flawed and even insulting picture of the "new man." Certainly, Chernyshevsky's heroes, unlike Turgenev's Bazarov, successfully resolve their love affairs and end, as the narrator promises, "happily, amidst wine and song" (Chernyshevsky, *What* 48). With that important difference in mind, their science, especially as practised by Lopukhov and (Alexander) Kirsanov, looks a great deal more like Bazarov's than not. Like Bazarov, Lopukhov and Kirsanov are trained as medical doctors, and as doctors with considerably less interest in treating actual patients than in conducting scientific research that again involves a great deal of dissection of frogs. As Chernyshevsky's narrator explains:

It's a curious thing: in the last ten years or so a number of our best medical students have decided upon graduation not to practice medicine – the only vocation that affords a medic the means of earning a decent living. At the first opportunity, they drop medicine and take up one of its auxiliary sciences – physiology, chemistry, or something of that sort. But every one of them knows that if he had a private practice, he would have a major reputation by the age of thirty, a secure life by the age of thirty-five, and by forty-five – great wealth. But they reason differently: you see, medicine is now in such an undeveloped state that, rather than treat patients, it's more important to prepare for the future so that doctors will possess the skill to administer treatment. And so, in the name of their beloved science (they delight in abusing medicine, even though they devote all their efforts to it), they reject wealth, even prosperity, and, you see, they sit in hospitals making interesting scientific observations. They dissect frogs, cut open hundreds of corpses a year, and at the first opportunity they outfit themselves with a chemical laboratory. (92)

Although Lopukhov eventually finds it in his own self-interest to leave the field, Kirsanov perseveres in the face of some resistance. When older university faculty attempt to withhold not only his chair, but even his degree, Kirsanov's students and even some of the younger professors begin to spread the idea that "there was a certain man named Virchow who lived in Berlin, and one named Claude Bernard who lived in Paris" (213–14). "Well," the narrator says, "it was this very same Claude Bernard who'd spoken respectfully about Kirsanov's work when the young student was still completing his studies" (214). With that seal of approval, Kirsanov is granted both his degree and his chair. Whether actually engaged in the practice of science or not, Chernyshevsky's heroes also talk a certain scientific talk, as does his narrator.

Bazarov, too, was a bit of a name-dropper, especially with regard to the (real) authors whose work he recommends: the German philosopher, physiologist, and physician Ludwig Büchner; German chemist Justus von Liebig (1803–73); and French chemists Théophile-Jules Pelouze (1807–67) and Edmond Frémy (1814–94). He displays only a limited command of the physiological jargon that litters the conversations of Chernyshevsky's heroes, however. Lopukhov and Kirsanov discuss their work on the "optical nerve" and the production of "artificial albumin," while Kirsanov in particular offers long, apparently scientific, disquisitions on the workings of sensation and the relationship of mind to body. "The intensity of sensation is in proportion to the level of feeling from which it evolves in the organism" (Chernyshevsky, *What* 358), he tells Vera Pavlovna, or: "Statistics have already demonstrated that the female organism is more resilient. You've read these conclusions only in life-expectancy tables. If you add physiological evidence to the statistical data, then the difference emerges as much greater" (340). Like Chernyshevsky himself in "The Anthropological Principle in Philosophy," the narrator of *What Is to Be Done?* is also explicit in his view that the scientific method can be extended from the laboratory into the wider sphere of human life, including the romantic entanglements that make up the plot of the novel. According to the narrator there is a "rule of logic" that he, too, "learned by heart": "The observation of phenomena that occur in and of themselves must be verified by experiments, conducted according to a well-formulated plan, in order to ensure the most profound insight into the mysteries of such relationships" (114).

Dostoevsky, starting with *Notes from Underground*, always takes aim at Chernyshevsky. As Chernyshevsky responded to *Fathers and Sons*, so Dostoevsky wrote *Notes from Underground* in answer to *What Is to Be Done?*, so much so that readers have long understood the "gentlemen" or "you" to whom the Underground Man addresses his tirade as referring at least in part to Chernyshevsky and his nihilist friends. Dostoevsky also makes a point of parodying different elements of Chernyshevsky's plot. Where in Chernyshevsky it is the rationally self-interested men who rescue women, and especially prostitutes, for example, in Dostoevsky it is the prostitutes who at least attempt to save the men; a careful reader might also note that where Lopukhov only pretends to commit suicide while really going to America, Svidrigailov in *Crime and Punishment* announces that he is going to America only to commit suicide. If Chernyshevsky evidently mattered enormously to Dostoevsky, still when Dostoevsky refers in his novels to actual science and scientists, he necessarily also reaches past Chernyshevsky to indict Chernyshevsky's sources. Russian nihilism, after all, as Kirsanov's training abroad would suggest, is not really Russian at all. It is rather a kind of science and also a way of being in the world that Western Europe knows as positivism.

Positivism

In the nineteenth century positivists, broadly speaking, are again a straight-forward, empirically grounded lot. "I'm a positive [положительный] man, an uninteresting one," Bazarov tells Odintsova: "I don't even know how to converse" (Turgenev 76, amended). Like Bazarov, Kirsanov, and Lopukhov, they are also practitioners of a particular kind of science, as in the authors of the real compendium of articles that Dostoevsky's Lebeziatnikov leaves with his friends the Kobyliatnikov ladies in *Crime and Punishment*: *The General Conclusion of the Positive Method* (*Общий вывод положительного метода*, 1866). While Lebeziatnikov recommends in particular "an article by Piederit [sic] (and, incidentally, one by Wagner as well)" (*Crime* 400; *PSS* 6: 307), the real luminaries in that collection were physiologists Claude Bernard and Jacob Moleschott and cell pathologist Rudolf Virchow.[2] Despite the title, and whatever one might say about the less well-known Theodor Piderit (1826–98) and Adolf Wagner (1835–1917), the former again a physiologist and the latter at least trained as an economist, it's fair to say that Bernard, Virchow, and Moleschott are equally "positive" only in a particular, and perhaps particularly loose, understanding of the term. While Bazarov's claims to an entirely unadorned presentation matching his matter-of-fact empiricism suggest "a positivism as divorced from politics and naively devoted to the project of a unified, disembodied, and ahistorical scientific language" (Tresch 256), as Tresch points out, that kind of positivism was, in fact, a twentieth-century invention, one put forward by Karl Popper and the Vienna Circle. Again, strictly speaking, nineteenth-century positivism as imagined by Auguste Comte was neither the twentieth-century version nor quite what the later nineteenth century thought it was. As Tresch argues in *The Romantic Machine*, positivism in its "original," Comtean form was a much stranger and much more "subjective" beast than even Comte's earliest advocates tended to remember, as was Comte himself, complete with his later invention of a religion of positivism as well as his periodic descents into actual insanity. It is nonetheless Comte that the Underground Man has in mind when he ridicules the notion that we need only "the laws of nature, the conclusions of natural science, mathematics" (*Notes* 13; *PSS* 5: 105) and "all human actions will then be calculated according to these laws, mathematically, like a table of logarithms, up to 108,000, and entered into a calendar" (24; *PSS* 5: 113), and also Comte, for all his contradictions, flaws, and even oddities, who echoes across nineteenth-century science.

Like Dostoevsky, Comte was trained in mathematics and the natural sciences, only much more briefly. Evidently something of a rabble-rouser, Comte was expelled from the prestigious École polytechnique in 1814 in only his second year as a result of his leading role in an attempt to oust an unpopular and royalist-leaning teacher. If Comte's actual education was shorter, however, his scientific ambitions lasted much longer: where Dostoevsky after his graduation from the Academy of Engineers served in the military only briefly before dropping engineering for a precarious living in literature, Comte aspired to a scientific career his entire life. After his abrupt departure from the École polytechnique, Comte found employment as a secretary to the utopian social reformer Henri de Saint-Simon (1760–1825). As Tresch recounts, Comte had already started work on what would become his magnum opus, *The Course on Positive Philosophy* (*Cours de Philosophie Positive*, 1830–42), when he attempted to re-join the scientific establishment with his first and only appearance at the Academy of Sciences in a claim to have mathematically proven the nebular hypothesis proposed by Pierre-Simon Laplace (1749–1827). The nebular hypothesis as it also appears in Edgar Allen Poe's "Murders in the Rue Morgue" (1841) argued that the planets were formed out of spinning clouds of very hot gas. In Tresch's summary, "as these 'nebulae' condensed, they left behind them planets that continued to revolve around the final kernel of condensation, the sun" (Tresch 266). Unfortunately, Comte's attempt to derive a correlation between the period of rotation of the supposed primary nebular mass and the orbits of the various planets was fundamentally flawed.

As Tresch writes: "To arrive at his figures for the successive revolutions of the nebular mass at the moment at which it would have left behind each planet, Comte had used equations that had themselves been derived from the periods of the current planets' revolutions." In other words, "Comte's logic was viciously circular" (Tresch 268), and while Comte's British supporters didn't learn of the mistake until 1844, in France it was exposed almost immediately. As time wore on, practising scientists like Claude Bernard would also take issue with the drive to systematization that for Comte often operated at the expense of empirically derived scientific facts. Many of his supporters were also not just mystified, but even actively embarrassed by Comte's turn by the late 1840s to the promulgation of his Religion of Humanity. If exercised accordingly from the margins, still Comte's influence on the broad contours of nineteenth-century scientific thinking is indisputable. Although, as Tresch writes, Comte was in fact far from disentangling *a priori* beliefs from empirical observations, Comte's broad claim to set aside "any research into what are called *causes*, whether first or final" (Comte 75), was enormously influential, including on Darwin. As again Tresch notes, "Darwin was exposed in Edinburgh in the 1830s not only to the transcendental anatomy of Geoffroy Saint-Hilaire, but to Comte's cosmogony; his eventual theory of evolution by natural selection had strong commonalities with both" (Tresch 268). In more specific terms, it was also Comte who generated tremendous enthusiasm across Europe for the extension of what he saw as the methods of science into new fields, including the newly minted "sociology," or, as Comte liked to call it, "social physics."

"Social physics" was perhaps especially appealing to non-scientists, including Chernyshevsky. Comte is mentioned only once by name in *What Is to Be Done?* and then not in entirely favourable terms when the energetic disputants at the jolly nihilist picnic toss around "Auguste Comtean" as an expression of apparently mild disparagement (Chernyshevsky, *What* 204). "The Anthropological Principle in Philosophy" would nonetheless suggest that Chernyshevsky fully embraced the essential Comtean notion of a hierarchy of the sciences. In Comte's view, as he wrote in the *Course*, there are five fundamental sciences "in successive dependence – astronomy, physics, chemistry, physiology, and finally social physics" (Comte 96), all arrayed under the government of mathematics taken, as Comte puts it, in its "abstract" or "instrumental" sense as "simply an extension of natural logic to a certain order of deduction" (100). Mathematics as the principle that takes in all the sciences is an aspiration to increasing scientific "hardness" that still marks Science with a capital S. As Chernyshevsky puts it in "The Anthropological Principle in Philosophy" in what Bethea and Thorstensson describe as his "confident, almost breezy" (Bethea and Thorstensson 38) style, it is only through its recourse to mathematics that a given science lifts itself out of intellectual poverty. "As soon as any gentleman, or lady, in a large family achieves a good position in society, he or she at once begins to drag his or her relatives out of poverty and obscurity," he explains. "Exactly the same thing is happening in the sphere of knowledge" (Chernyshevskii, *Selected* 90). As pleased as Chernyshevsky is to see chemistry borrow from mathematics to reach the level of perfection that celestial and Newtonian mechanics already enjoy, what thrills him most along with positivists everywhere are the possibilities opening up for the sciences "concerned with plant and animal organism": physiology, comparative anatomy, various branches of botany and zoology, and especially "the moral sciences."

"It is not so long ago," Chernyshevsky asserts with some pride:

> that the moral sciences could not have possessed the content that could justify the title of science that they bore, and the English were quite right then in depriving them of a title they did not deserve. The situation today has changed considerably. The natural sciences have already developed to such an extent that they provide much material for the exact solution of moral problems too. (Chernyshevskii 92)

In *Crime and Punishment*, Raskolnikov's idea is exactly the scientific solution of moral problems, both in the plan to kill the old pawnbroker that he supposes is "true as arithmetic" (*Crime* 59; *PSS* 6: 50) and when he attempts to convince himself that a young girl's abuse and degradation are inevitable. "They say that's just how it ought to be," Raskolnikov thinks to himself, "Every year they say, a certain percentage has to go … somewhere" (50; *PSS* 6: 43). Like Chernyshevsky's, Raskolnikov's vague recourse to statistical thinking also reflects not just Comte, but another scientist influenced by Comtean thought: Belgian mathematician and social scientist Adolphe Quetelet.

While Quetelet started as an astronomer and even studied briefly with Laplace, it was also Laplace who shifted his focus from celestial mechanics to probability and social statistics. In 1835, Quetelet published his two-volume *Sur l'homme et le développment de ses facultés, ou essai de physique sociale* (translated into English in 1842 as *A Treatise on Man and the Development of His Faculties*, and in Russian as both *Человек и развитие его способностей, или Опыт общественной физики* and *Социальная физика, или Опыт исследования о развитии человеческих способностей* only in 1865); Quetelet himself usually referred to it in Comtean terms as his *Physique sociale*. As Stephen Stigler writes, "There was no mistaking Quetelet's aim in this book: to lay the groundwork for a social physics, to conduct a rigorous, quantified investigation of the laws of society that might someday stand with astronomers' achievements of the previous century" (Stigler 170). As Stigler notes more than once, Quetelet's initial data was often taken in error, he tended to account for anomalies in ways that suited his own preconceived notions, and his work even "abounds in numerical errors, an indication that he frequently calculated in haste and lacked the patience to recheck his work" (180). Still, his invention of the "average man," "*l'homme moyen*," first as a means of revealing the laws of "social physics" and then, in a process of increasing materialization, as a "type" to which nature aspired, proved extremely influential. As Stigler says, "The average man was a fictional being in his creator's eye, but such was his appeal that he underwent a transformation, like Pinocchio or Pygmalion's statue, so that he still lives in headlines of our daily papers" (170). It is also Quetelet's work that then lent itself to Henry Thomas Buckle's still more magisterial attempt to identify the laws of history in his *History of Civilization in England* (1857).

In *Notes from Underground* Dostoevsky takes something of a sideswipe at Buckle when his narrator describes the "logistics" of rational self-interest as "almost the same as … well, let's say, for example, the same as asserting, with Buckle, that man gets softer from civilization and, consequently, becomes less bloodthirsty and less capable of war" (*Notes* 23; *PSS* 5: 110). The real Buckle was a great admirer of Quetelet whose "social physics" inspired his own "science of history" (Buckle 1: 18). As Buckle writes:

> M. Quetelet, who has spent his life in collecting and methodizing the statistics of different countries, states, as the result of his laborious researches, that "in every thing which concerns crime, the same numbers re-occur with a constancy which cannot be mistaken; and that this is the case even with those crimes which seem quite independent of human foresight, such, for instance, as murders, which are generally committed after quarrels arising from circumstances apparently casual." (1: 24)

It is Buckle's claim that his own work is based on "exclusively inductive" (1: 21) investigations founded in turn "on collections of almost innumerable facts, extending over many countries, thrown into the clearest of all forms, the form of arithmetical tables" (1: 21). The scientific historian, he argues, can only work from a concept of law: "Rejecting, then, the metaphysical dogma of free will, and the theological dogma of predestined events, we are driven to the conclusion that the actions of men, being determined solely by their antecedents, must have a character of uniformity, that is to say, must, under precisely the same circumstances, always issue in precisely the same results" (1: 19–20). While Buckle, like Quetelet, was read across Europe, readers even at the time took issue with an ideological stance only barely masquerading as empirical science. Concurrent developments in the biological sciences enjoyed more universal respect at the time and still do today. Still, the expansion of what Chernyshevsky calls the "union of the exact sciences" (Chernyshevskii, *Selected* 91) into the life sciences lent itself to a similar set of problems.

Of the contributors to Lebeziatnikov's *The General Conclusion of the Positive Method*, Claude Bernard is best known as a sometime Comtean, although by 1865 and his widely read *Introduction to the Study of Experimental Medicine* Bernard was at pains to distance himself from his Comtean beginnings. Bernard is still very highly regarded for his concept of *milieu intérieur*, a precursor of what scientists today call homeostasis, as well as for his development of a scientific method that he adhered to far more rigorously in his laboratory than did Buckle, Quetelet, or Comte in the absence of any actual experimental work at all. In his *Introduction* Bernard with his focus on both experimental intervention and "objective," empirical observation openly derides "a modern philosophic school which piques itself on its scientific basis" but "engender[s] nothing but systems" (Bernard 25); he is also quite wary of the notion that we might indiscriminately apply the tools of mathematical analysis and especially statistics to other, more "real" phenomena. If Bernard makes his caveats clear, however, his Comtean starting point is equally evident.

Bernard takes directly from Comte, for example, his notion of a three-fold development of human thought. For Comte "each of our leading conceptions – each branch of our knowledge – passes successively through three different theoretical conditions: the theological, or fictitious; the metaphysical, or abstract; and the scientific, or positive" (Comte 71). Bernard is after the same idea when he describes "the experimental method" as "by no means primitive or natural to man," but achieved "only after lengthy wanderings in theological and scholastic discussion" (Bernard 27). He is also at his most Comtean when he claims that what the experimental sciences reveal is that "primary causes, like the objective reality of things, will be hidden from [man] forever, and that he can know only relations" (28). Bernard also aspires exactly to the "unified, disembodied, and ahistorical scientific language" that Comte promised but failed to deliver. "The experimental method is concerned only with the search for objective truths, not with any search for subjective truths" (28), Bernard writes. While his professional self tries to keep his goals modest, we finally catch glimpses also of an underlying arrogance.

Bernard's laboratory experiments depend on a practice of "dissociation," or vivisection, that he knows that he needs not just to explain, but also to justify. "If a comparison were required to express my idea of the science of life," he writes, "I should say that it is a superb and dazzlingly lighted hall which may be reached only by passing through a long and ghastly kitchen" (Bernard 15). As we might guess, the dissection of frogs plays a prominent role in that passage. "If the frog," Bernard notes, "as has been said, is the Job of physiology, that is to say, the animal most maltreated by experimenters, it is certainly the animal most closely associated with their labors and their scientific glory" (115). Bernard's own work nonetheless relied largely on the vivisection of rabbits and dogs instead, including in his experimental determination that white chyle (a bodily fluid consisting of lymph and emulsified fats) is formed by the pancreatic juices (154). As Bernard was well aware, as increasingly widespread as the practice of vivisection was in the nineteenth century, and even as well-regarded as its discoveries were, still many of his contemporaries were actively horrified by that "long and ghastly kitchen." Wilkie Collins's late novel *Heart and Science* (1883), for example, draws on the anti-vivisectionist argument put forward by Lewis Carroll (1832–98) in his 1875 pamphlet "Some Popular Fallacies about Vivisection" and features a vivisectionist scientist as its villain.[3] In France, Bernard's own wife not only left him in the apparent belief that he "would have done better to apply his energies toward a prosperous medical practice instead of tormenting helpless animals in his laboratory" (Virtanen 2), but after his death contributed money to anti-vivisectionist societies in hopes that she could atone for what another contemporary called "the crimes of experimental physiology" (119).[4] Bernard defended his science exactly as Bakhtin's "mastery over mute objects, brute things," however, and "mastery" is putting it mildly. Bernard tries to rein in his ambitions as he does any "preconceived ideas": "The truly scientific spirit," he writes, "should make us modest and kindly" (Bernard 39). Bernard was nonetheless anything but kindly to the frogs and rabbits under his care, and he also occasionally lets slip what can only be called a vaunting ambition. "With the help of these active experimental sciences," he exults, "man becomes an inventor of phenomena, a real foreman of creation; and under this head we cannot set limits to the power that he may gain over nature through future progress in the experimental sciences" (18).

While the passage of time has complicated our view of Thompson's "heroic materialism," still much of Bernard's science retains its real sophistication, as does Virchow's pioneering work in cell pathology. The same cannot be said, however, for their fellow contributor to *The General Conclusion of the Positive Method*, Jakob Moleschott. Moleschott gained notoriety together with Ludwig Büchner and Karl Vogt as a proponent of what Friedrich Engels (1820–95) famously mocked as "vulgar itinerant-preacher materialism" (Engels 340). For Engels their materialism was "vulgar" in that their radically monistic insistence on a world of matter alone failed to anticipate the Marxist view that the tangible world is only a symbol of the "real" reality of economic relationships. For the non-Marxist, the three were "vulgar" in that they were extremely popular. In Frederick Gregory's count, Büchner's 1853 *Matter and Force* (*Kraft und Stoff*) went through twelve editions in seventeen years and was translated into seventeen foreign languages, including Russian; when Arkady's father spends too much time reading Pushkin in *Fathers and Sons*, Bazarov suggests that Arkady give him Büchner instead.[5] Vogt is perhaps better remembered today than Büchner, but only for a single line, his inflammatory claim in 1846 that "thoughts stand in the same relation to the brain as gall does to the liver or urine to the kidneys" (Gregory 64). While among the three it was Moleschott who possessed the most in the way of actual scientific credentials, still Moleschott's works, like Vogt's and Büchner's, were aimed at a broad, general audience, and Pisarev's popular review of what was already a work of popular science, Moleschott's relatively late *Physiological Sketchbook* (*Physiologisches Skizzenbuch*, 1861), effectively illustrates his more than simplistic approach.

In his review of the *Physiological Sketchbook* Pisarev makes a point of offering his reader an array of facts gleaned from his reading. "[B]lood is made up of a combination of nitrogen, carbon, hydrogen, oxygen, potassium, sodium, calcium, magnesium, iron, sulphur, phosphorus, chlorine and fluorine" (Pisarev 3:155), he explains; or: "In raw meat the meat fibres are surrounded by a sort of juice consisting of a solution of protein, various salts, and nitrogeneous creatine (Fleischstoff)" (3: 162). For all the complicated pseudo-scientific jargon that the graduate in philology clearly enjoys, the main thrust of Moleschott's argument in the *Physiological Sketchbook*, as elsewhere, is highly reductive. Where Marx's base is a more complicated but also less material set of economic relations, Moleschott's is purportedly matter alone: as Feuerbach famously wrote in his review of Moleschott's earlier *Die Lehre der Nahrungsmittel: Für das Volk* (1850; translated into English in 1856 as *The chemistry of food and diet, with a chapter on food adulterations*), "Der Mensch ist was er ißt," "Man is what he eats" (Cherno 399). In his article on the *Sketchbook* Pisarev even quotes from *Die Lehre* to hammer this point home. "Can lazy potato blood possibly lend muscles the strength for work and impart to the brain the life-creating impulse of hope?" his Moleschott cries. "Poor Ireland! Your poverty gives birth to poverty! You cannot remain unconquered in the struggle with a proud neighbor to whom plentiful herds impart power and boldness!" (3: 158). As Pisarev celebrates it, "vulgar" materialism reveals every aspect of our reality as nothing more than a predetermined effect of matter.

Minds and Bodies in the World

Pisarev is not alone in his enthusiasm for the simple verities of the "vulgar" approach; indeed, "potato blood" had and has a great deal of appeal. Still, material determinism together with an all-encompassing "algebraicization" of the natural and even the social world combined with a scientist as "a real foreman of creation" makes for a strange sort of science, one that for all its popularity is not really very scientific at all. It is also one that wasn't widely advocated by actually practising nineteenth-century scientists, including not just Darwin, Lewes, and Helmholtz, but also the most famous of nineteenth-century physicists, James Clerk Maxwell (1831–79). Among non-scientists, Maxwell is perhaps best known for the thought experiment that his friend and fellow physicist William Thomson called "Maxwell's demon." Maxwell's "demon" was an imaginary creature stationed at a door separating the two halves of a gas-filled container and endowed with the ability to sort fast gas molecules from slow and so contravene the second law of thermodynamics (the necessary increase in entropy in a closed system). As Maxwell described it in a letter to another friend:

Concerning Demons.

1. Who gave them this name? Thomson.
2. What were they by nature? Very small BUT lively beings incapable of doing work but able to open and shut valves which move without friction or inertia.
3. What was their chief end? To show that the 2nd law of Thermodynamics has only a statistical certainty. (I. Tolstoy 136–7)

The layperson more comfortable with the simpler science of the nihilist/positivist sort may struggle to reconcile Maxwell's insistence here on the ultimate uncertainty of the particular and the contingent with his great claim to fame in the scientific world, his extraordinary discovery of the equations that describe the phenomena of electromagnetism, including light. For all the capacity of mathematics to depict it, though, still reality for Maxwell was made up of concrete, individual instances and entities, and Quetelet was in another business altogether.

It's not that a study of "the character and propensities of an imaginary being called the Mean Man" is entirely without value, Maxwell wrote, but that what he calls "exact science" (Campbell 440) is an entirely different endeavour, one that aspires not to invent imaginary beings but to adequately describe the concrete individuals and instances that make up the real world. As Jane Bennett has more recently argued, in the absence of the particular, the contingent, and the world as it really is, "vulgar" materialism loses sight of the very material world it would purport to consider. For Maxwell, as for Lewes and Helmholtz, the latter both a physicist like Maxwell and a physiologist like Lewes, a true engagement with the material world would finally also entail an acknowledgment of mind as not separate from the world that it would consider. Just as Bruno Latour in our own day insists that mind is itself "a wriggling and squiggling part of nature" (Latour 10), so for Maxwell "the only laws of matter are those which our minds must fabricate, and the only laws of mind are fabricated for it by matter" (I. Tolstoy 77). Dostoevsky encountered this other sort of nineteenth-century science in his life-long reading of leading European scientists and scientifically inclined European novelists, all of whom appeared in the Russian press, as in the scientifico-literary circles that he ran in at home in Russia. His investment in minds and bodies as twin aspects of what Vadim Shneyder memorably calls a "stubborn" [прочная] materiality (Shneyder 77) was finally also shaped by his reading of a philosopher who once upon a time took Europe by storm: Ludwig Feuerbach.

Feuerbach's *The Essence of Christianity* (*Das Wesen des Christentums*, 1841) reached Russia in January 1842. Pavel Annenkov remembered it as "in everybody's hands" by the mid-1840s. "It can safely be affirmed," he wrote, "that Feuerbach's book nowhere produced so powerful an impression as in our 'Western' circle, and nowhere did it so rapidly obliterate the remnants of all preceding outlooks" (J. Frank 186). In 1853 in far-away England, Marian Evans, not yet George Eliot, was so taken with her reading of Feuerbach's blockbuster that she published her own translation, one that remains in print today, while Friedrich Engels remembered a more general pandemonium: "One must himself have experienced the liberating effect of this book to get an idea of it. Enthusiasm was general; we all became at once Feuerbachians" (Feuerbach, *Fiery* 2). Back in Russia, Chernyshevsky read Feuerbach at the age of twenty-one and even described himself as a "devout Feuer-bachian" (Paperno 198). Dostoevsky read Feuerbach at the same age and in the same left-wing political context that for Dostoevsky came to an abrupt end with his arrest in 1848 together with the other members of the Petrashevsky "circle." To associate Feuerbach with radical politics alone, however, is to diminish the real force and also range of his thought. Certainly, Chernyshevsky claimed him, as did Marx, although in the case of the latter only to later dismiss him; Feuerbach is also most well-known now as he was in his own day for the single sentence from the review of Moleschott quoted above: "Der Mensch ist was er ißt." If a fully adequate summary of Moleschott's theory of "potato blood," however, this most famous line, as Melvin Cherno has argued, finds no resonance in Feuerbach's writings as a whole.[6] By the same token, for all the inspiration that he offered to the purveyors of both historical and "vulgar" materialism alike, Feuerbach's own materialism operated in strikingly different terms.[7]

As Aileen Kelly notes, Alexander Herzen was yet another of those young people reading Feuerbach in the 1840s, in his case as a student in the Faculty of Physical and Mathematical Science at Moscow University. In Kelly's reading, this early grounding in the sciences is the key to Herzen's later thought, as Herzen's philosophy of history is most importantly marked by a deeply Darwinian "theme of contingency" (Kelly 6). "In the same year as Marx's pronouncement that communism was 'the solution of the riddle of history,'" Kelly writes, "Herzen declared that there were 'no solutions': history, like nature, was an improvisation, subject to the play of chance" (5). At heart always a natural scientist, Herzen in Kelly's account was then also shaped by his reading not just of Feuerbach, but of Feuerbach together with German philosopher, poet, and playwright Friedrich Schiller. In associating two such apparently "strange bedfellows," Kelly emphasizes Schiller's own background in the sciences, both in his original training as a medical doctor and in his early scientific writing that, in Kelly's words, "reflect[s] the empirical approach to the relation between body and mind then being pioneered by such figures as the Swiss physiologist Albrecht von Haller" (222). The same attention to the material functioning of real minds and bodies in the world marks Feuerbach's thought as well.

Feuerbachian materialism in *The Essence of Christianity*, as in the more aphoristic *Principles of the Philosophy of the Future* (*Grundsätze der Philosophie der Zukunft*, 1843), is predicated on the centrality of what Feuerbach again and again calls "sensuousness." As he explains in the latter work, the philosophy of the past sought "to avoid sensuous conceptions ... in order not to pollute abstract concepts." In contrast, Feuerbach writes, his new philosophy "*joyfully* and *consciously* recognizes the truth of sensuousness: It is a *sensuous* philosophy with an *open heart*" (Feuerbach, *Fiery* 227). "Only to an open mind does the world stand open,"

Feuerbach adds, "and the *openings of the mind* are only the senses" (241), even when it comes to man's knowledge of himself:

> We feel not only stones and wood, not only flesh and bones, but also feelings when we press the hands or lips of a feeling being; we perceive through our ears not only the murmur of water and the rustle of leaves, but also the soulful voice of love and wisdom; we see not only mirror-like surfaces and specters of color, but we also gaze into the gaze of man. (230)

Feuerbach's is not a material world devoid of thought, but a world where thought as both imagination and reason is itself always embodied. In *The Essence of Christianity* Feuerbach describes the imagination as the "limitless activity of the senses" (Feuerbach, *Essence* 214); in *Principles of the Philosophy of the Future* he explains that the new philosophy "is certainly based on reason as well, but on a reason whose *being* is the same as the *being of man*; that is, it is based not on an empty, colorless, nameless reason, but on a reason that is *of the very blood of man*" (Feuerbach, *Fiery* 239). As should already be evident, the italics are typical of Feuerbach's style, as is the repeated recourse to often visceral material images.

Feuerbach writes not just of the *"very blood of man,"* but of the "systole and diastole" of life that finds its most concrete manifestation as "the activity of the arteries drives the blood into the extremities, and the action of the veins leads it back again" (Feuerbach, *Fiery* 129). He also compares a would-be subject lacking the object that would complete it to the actual physical discomfort of hunger pangs: "[t]he pain of hunger means that there is nothing objective inside the stomach, that the stomach is, so to speak, its own object, that its empty walls grind against each other instead of grinding some content" (226). Paul Bishop notes the frequent appearance of animals in Feuerbach's philosophical writings: a flea, a caterpillar, a frog, a pig, even the bird to whom God can only be a winged being because, as Feuerbach explains in his introduction to *The Essence of Christianity*, the "highest being to the bird is the 'bird-being'" (Feuerbach, *Fiery* 114). The repeated use of materialist imagery serves to underscore the fundamental emphasis on "sensuousness" that is also Dostoevsky's, from the Underground Man's "living life" through the "sensuality" that marks all the Karamazovs, from their "sensualist" (*Brothers* 8; *PSS* 14: 8) of a father to Dmitri's "insect sensuality" (109; *PSS* 14: 99) and Ivan's sticky little leaves in spring. Feuerbach's many examples from the physical or even non-human world also operate to express the full dimensions of his commitment to the dynamic interrelationship of self and other.

Feuerbach's fundamental argument in *The Essence of Christianity* is that religion, like speculative philosophy, cuts off from the real world to worship an object of its own invention, a projection of what is best in man himself: "The divine being," he writes, "is nothing else than the human being, or, rather the human nature ... made objective – i.e. contemplated and revered as another, a distinct being" (Feuerbach, *Essence* 14). Feuerbach himself claimed that his intent was to cast out only the form of Christianity, not its content. As he put it in some exasperation in the preface to the second edition of *The Essence of Christianity*: "it is to be hoped that my readers, provided that they are *not stone-blind*, will realize and be led to the conviction – even if reluctantly – that my work represents an *authentic translation* of the Christian religion from the original language of images into straightforward, intelligible German" (Feuerbach, *Fiery* 251). Dostoevsky famously didn't see it that way, and his fierce response to what for many readers were the further implications of "anthropological Christianity" is well-known: for Dostoevsky the philosophy of the man-God leads not to life, but to death. If Dostoevsky emphatically rejected any hint of atheism, still, as A.A. Kazakov argues, Feuerbach's underlying notion of the mutual creating of self and other may well have served as a source for his own notion of dialogism. Certainly, Feuerbach inspired George Eliot's similar emphasis on what the latter calls fellowship.

As Beer notes, it was only after her translation of Feuerbach that George Eliot emerged as a novelist. Eliot, she writes, "found in Feuerbach's emphasis on awakened imagination ... an idea that helped to liberate her own creativity" (Beer, *George Eliot* 75). In Beer's account, the imagination conceived as the "limitless activity of the senses" then led Eliot not just to write, but to understand her writing in a particular way, as a "joint enterprise of production" that "'furnishes space' for the activity of reader and writer together" (75–6). The Feuerbachian relationship of self and other finally leads to a principle of love that served Eliot as philosophical justification for her own unconventional life choices. "Love is God himself, and apart from it there is no God," Feuerbach exults in Eliot's own translation, "not a visionary, imaginary love – no! a real love, a love which has flesh and blood, which vibrates as an almighty force through all living" (Feuerbach, *Essence* 47). A sensual love in the world is love for a real other who is at the same time an extension and a completion of the self, and Feuerbach presents this love in distinctly scientific as well as typically emotional terms. "A loving heart is the heart of the species throbbing in the individual," Feuerbach writes. "Thus Christ, as the *consciousness of love*, is the *consciousness of the species*" (269). For Dostoevsky, Christ is not a reflection of ourselves, but he is love, and love in the real world, as in the material, sensual love that Prince Myshkin in *The Idiot* (*Идиот*, 1868) can't quite achieve, or in the kiss that Alyosha Karamazov gives his brother Ivan. In Dostoevsky, too, if Christ is in a special category, still we are all a reflection of one another, a vibrating whole of multiple parts that, as in Feuerbach, expresses also our relationship to the material world.

Just as Hermann von Helmholtz in a more strictly scientific vein explored the mechanics of sound only ever in terms of the receiving capacity of the ear, so Feuerbach in *Principles of the Philosophy of the Future* asks: "What is light – as the shining and illuminating being, as the object of optics – without the eye?" (Feuerbach, *Fiery* 199). Feuerbach also makes a point of extending this sense of the mutually constituting activity of subject and object to the apparently inanimate world. The sun functions for the planets, Feuerbach argues, much as God functions for man, that is to say, as any object functions for any subject. In *The Essence of Christianity* he writes:

> Each planet has its own sun. The sun which lights and warms Uranus – and the way it does so – has no physical (only an astronomic or scientific) existence for the Earth. Not only does the sun appear different, but it really is another sun on Uranus than on the Earth. Hence, Earth's relationship to the sun is at the same time the Earth's relationship to itself, to its own being, for the measure of the magnitude and intensity of light which is decisive as to the way the sun is an object for the earth is also the measure of the Earth's distance from the sun, that is, the measure that determines the nature of the Earth. The sun is therefore the mirror in which the being of each planet is reflected. (Feuerbach, *Fiery* 101)

The mutually informing activity of self and other is everywhere in Dostoevsky, in the dense interrelationship of Raskolnikov and his room or Raskolnikov and his city, for example, as in Raskolnikov's resurrection from the dead only when he recognizes his love for Sonia. It also served as a guiding principle in the work of Dostoevsky's sometime friend and long-time collaborator, Nikolai Strakhov.

Strakhov was yet another seminarian who came to St. Petersburg to study, in his case not philology, but, remarkably, first law, then mathematics, and finally zoology. Strakhov also taught the natural sciences at the secondary level for more than ten years before making a career as a contributor for a number of the journals of the 1860s and 1870s, including most notably the Dostoevsky brothers' journals *Time* (*Время*, 1861–3) and *Epoch* (*Эпоха*, 1864–5). At *Time* Strakhov, together with the Dostoevskys and their leading literary critic, the brilliant if erratic

Apollon Grigoriev (1822–64), advocated the science-inflected theory of "pochvennichestvo," an "organic criticism" that called for a return to the "soil" [почва] in vaguely biological terms, ranging from "vegetative poetry" and "antediluvian formation" to "drift" [веяние] and "locality." The Grigorievian claim that the "organic view" takes "creative, direct, natural life forces as its point of departure – in other words, not the intellect alone ... but the intellect and its logical demands *plus* life and its organic phenomena" (Whittaker 416) exactly echoes Feuerbachian "sensuousness." As Zawar Hanfi writes, the great appeal of Feuerbach's philosophy was "the reunion it proclaimed of a nature released from its confinement to a dark, inferior, truthless order of reality to be overcome and transcended by an antithetical reason, and man as a being of plentitude capable of pouring himself into the infinite richness of religion, art, and philosophy" (Feuerbach, *Fiery* 20). The combination of "the intellect and its logical demands *plus* life and its organic phenomena" also drove Strakhov's more explicitly scientific work, culminating in his 1872 collection *The World as a Whole* (*Мир как целое*).

Strakhov's very real background in the sciences lent itself to a highly idiosyncratic approach. Strakhov consistently described himself as neither an idealist nor a materialist, and while he was strongly anti-Darwin, he nonetheless believed in his own sort of evolution. As L.P. Avdeeva writes, Strakhov argued "that organisms should be understood not as bodies, but as processes" wherein any given characteristic is not "a constant or fixed attribute, but instead appears and disappears" (Avdeeva 129). Like Maxwell with his demon, Strakhov also advocated experimentation while insisting on its descriptive, rather than prescriptive, value. "Chemists have thought that there is an absolute measurement for simple bodies and that is an easy matter for science to hit upon that measure," Strakhov writes, "[b]ut experiment doesn't produce an absolute" (Strakhov 446). Strakhov was finally intensely aware of our minds as a living part of the material world that we would consider.

Science itself bears the imprint of human intellect as Strakhov argues in his 1869 book *On the Method of the Natural Sciences and Their Significance in General Education* (*О методе естественных наук и значении их в общем образовании*) when he qualifies any possible notion of objectivity in strikingly modern terms. "Chief in science," he writes, "is the thought with which we look on nature, the goal towards which we strive. An important fact means that it is important for that thought and that goal; a surprising fact means that it doesn't fit with our thought; a negligible fact means that we don't see its meaning. Accordingly, the essential part of science lies not in facts, but in our view of facts" (Strakhov 5). In *The World as a Whole* he puts the interaction of subject and object not in terms of planet and sun, but man and material world. "Man is the light that illuminates the world," Strakhov writes, "and one can also say the opposite, that the world for each man is that sphere which is illuminated by the light of his consciousness" (Strakhov 201). Strakhov's science, I would argue, is also Dostoevsky's, indeed, as Dostoevsky once wrote to Strakhov, "You know, a good half of my views are your views" (Strakhov 24). Like Strakhov, Dostoevsky expressed that science in his own contributions to the literary theory of "pochvennichestvo." Like a great many other nineteenth-century writers, from Eliot and Schiller to Collins, Tolstoy, and Nathaniel Hawthorne, he also expressed that science in his novels.

In his 1961 "Text on Electricity" ("Texte sur L'Électricité") Francis Ponge (1899–1988) calls for a new kind of art that would reflect a new era and the "style of life" that has been ours "since the electrical current was placed at our disposal" (Ponge 157). As Ponge plays with the idea of a light that goes on and off at the flick of a switch by abruptly stating the intent of his piece and equally abruptly moving off on apparent tangents, the "Text on Electricity" itself demonstrates just what an "electric" literature might entail. When it comes to embodying the further revelations of science post-Einstein, however, including "the principle of uncertainty, and the relativity of Space and Time, and … the hypothesis of the indefinite extension of the universe" (179–81), Ponge can only propose that we start by clearing out the detritus of the past. "Our forms of thought, our rhetorical figures, actually date from Euclid," Ponge writes: "ellipses, hyperboles, paraboles, are *also* figures of that geometry." If we're to invent anything new, Ponge argues, we must first melt our now obsolete rhetorical figures "back into a mass, as it is done with old statues, in order to make cannons of them": "[t]hus, we may perhaps, one day, create new Figures that will allow us to put our trust in the Word, in order to traverse curved Space, non-Euclidean Space" (185). Before any aspiring twentieth- or twenty-first-century writers heed Ponge's call, I would suggest that we first take a good look at what we already have, starting with Dostoevsky. Nineteenth-century science isn't quite as Euclidean as Ponge seems to think, nor is it quite as definite. The nineteenth-century novel isn't, either.

2 Of Doctors and Detectives, Chemistry and Mathematics

In his 1961 "Text on Electricity," Ponge's call for a post-Euclidean art is pointedly cast as a matter of "rhetorical figures," form and not content. By the same token, the "Text" itself doesn't offer a plot, or even much in the way of characters: there's only an "I" who seems to correspond roughly to Ponge himself, along with a "you" who includes the presumably real architect-readers for whom the text was originally commissioned. Evidently for Ponge, as for many a modernist, "forms of thought" produce their own content, and the reliance on plot and character that marks the nineteenth-century novel is itself an indication of an outmoded way of writing. As we will note in the next chapter, the functioning of plot as a literary device in the nineteenth-century novel in exactly scientific terms is often far more interesting and certainly more forward-thinking than Ponge's practice would imply. This chapter will argue that actual content also matters. Science and scientists, both fictional and real, make frequent appearances in nineteenth-century literature, including in Dostoevsky's novels. Those references don't always serve the same epistemological ends, however.

We are perhaps most familiar with the nineteenth-century appeal to the sort of science that Thompson calls "heroic materialism," one that would "solve the world's problems 'once and for all'" (Thompson 192). Chernyshevsky offers an obvious example with his tale of happy nihilist scientists, including not just Lopukhov and Kirsanov, but also Vera Pavlovna, who by the novel's end has followed her two husbands' lead and herself embarked on the study of medicine. Chernyshevsky's "heroically materialist" content also comprises not just what his heroes do, but what they say. Chernyshevsky's "new" men and women all pepper their conversations with reference to the latest in science, from current research on the optical nerve to the production of artificial albumin; Vera Pavlovna even dreams of Liebig and the chemical composition of soil. The science that the nineteenth-century novel celebrates, however, isn't always of the "heroic" nihilist/positivist sort.

Middlemarch, for example, operates in terms of a double time frame, the 1860s–70s of its writing and the 1820s–30s of its setting, the latter admittedly a little too early for nihilist/positivist science as I have defined it.[1] As Eliot's readers we nonetheless recognize the markers in Lydgate, who, when we meet him, is only recently returned from studies in France where he engaged in "galvanic experiments" with rabbits and frogs. Now back in England, his ultimate ambition is to pursue not patients, but research, as he is "ambitious above all to contribute towards enlarging the scientific, rational basis of his profession" (Eliot, *Middlemarch* 147). Although not for Eliot's characters in their historical setting, the failure first of Lydgate's marriage and then of his career marks the failure of the Chernyshevskyan positivist/nihilist project, even as Eliot's narrator argues explicitly for another sort of science, one that acknowledges the mutual implication of mind and body and mind and material world. While it's not the traditional reading, this other, more complicated and still cutting-edge approach to the material world is also at work in the many references to science and scientists that mark the new genre of detective novel.

As a great many scholars post-Foucault rightly note, the detective novel after Poe's 1841 "Murders in the Rue Morgue" emerges alongside a science of criminology that by the end of the century had developed the early tools of forensic science, including finger-printing, the lie detector, and Bertillon's system of anthropometry, in Russia as in the West.[2] The nineteenth-century science of detecting has accordingly most often been read as a tale of Latour's "Science with a capital S," an almost magical combination of radical rationalism and equally rigorous empiricism that solves all mysteries exactly as Thompson describes. As Vanessa L. Ryan writes:

The association of the detective with superior powers of observation, vast scientific and human knowledge, and, above all, the use of scientific method, has made systematic thinking seem indispensable to the detective's art. In the world of Sherlock Holmes – the master of deductive logic and forensic analysis – the figure of the detective tends to correspond to our ideal of the pure scientist (Ryan 29).

The story of science that detective fiction tells can't always be read in those terms, however, even in the extreme case of Sherlock Holmes. To read Holmes as a "systematic" thinker, for example, is to gloss over the implications of his addiction to cocaine, while, in Thomas Sebeok and

Jean Umiker-Sebeok's account, Holmes even in his "normal" state relies not on deduction and "scientific method," but on a kind of inspired guessing that the Sebeoks call, after Peirce, "abduction." Even Poe as a pioneer of the form doesn't quite fit the mould that he is held to have invented, as where Michael Holquist claims that Poe's detective Dupin serves as an "instrument of pure logic" (Holquist 156), Albert D. Hutter finds instead a "relentlessly logical process of ratiocination ... thrown into question by a deeper irrationality" (Hutter 191). It is finally only a science of a less "systemic" sort that solves the mysteries at the centre of the nineteenth-century detective fiction closest to Dostoevsky's own: the novels of Wilkie Collins.

Collins is closest to Dostoevsky first in entirely tangible terms, as their work often appeared side-by-side in the pages of Mikhail Katkov's (1818–87) *Russian Herald* (*Русский вестник*). Over the course of 1866, for example, *The Russian Herald* serialized Collins's *Armadale* (1866) together with *Crime and Punishment*; in 1868, when Dostoevsky's *Idiot* was being serialized in the main part of the journal, it was with *The Moonstone* (1868) in the supplement [приложение]. This proximity is also a shared response to the limitations of nihilist/positivist science. In Collins, "heroic materialism" is the science practised by the villains, including the sublimely evil Count Fosco in *The Woman in White*, as well as Mrs. Lecount in *No Name* (1862), widow of the famous Swiss naturalist and current caretaker of his reptiles, and finally Dr. Benjulia, the Claude Bernard–inspired vivisectionist in *Heart and Science* (1883). It is left to Ezra Jennings in *The Moonstone*, once a doctor and now a disgraced doctor's assistant, to demonstrate a better kind of science in a "bold experiment" that solves the mystery of the missing diamond only by complicating the very idea of crime and criminal intentions. Dostoevsky's novels, for all their detecting, offer little in the way of hero-doctors, disgraced or otherwise, and his lone professional detective in *Crime and Punishment* is largely lacking in the science department. When Dostoevsky himself complicates the very idea of crime and criminal intention in *The Brothers Karamazov*, however, he does so most importantly with reference to a different but still more radical advance in nineteenth-century science: non-Euclidean geometry.

The topic of non-Euclidean geometry is broached by Ivan Karamazov, not that he is actually for it. As again Thompson points out, Ivan in Dostoevsky's last novel is his "first hero-scientist" (Thompson 205), apparently the only character in all of Dostoevsky to enjoy a formal training in the sciences. The narrator's very brief mention that Ivan "had graduated in natural science" (*Brothers* 16; *PSS* 14: 16) would seem at first glance an obvious marker of his nihilist tendencies, and, indeed, Ivan in his later conversation is clearly uncomfortable with the radically post- or anti-nihilist/positivist idea of different geometries to suit different kinds of spaces. That Ivan even entertains the possibility, however, speaks to the puzzle that Ivan is himself. Just as non-Euclidean geometry derives from our realization that even the terms of mathematics lend themselves to more than one meaning, so an education in the sciences, as Ivan himself eventually proves, needn't point only one way. If, as it turns out, there is more than one possible answer to the puzzle that is Ivan's personality, so is there finally more than one solution to the murder mystery that makes up Dostoevsky's plot. Who killed Ivan's father Fyodor Pavlovich? It depends on what you mean.

The Science of "Extraordinariness"

The mention of Ivan's actual training in the sciences in *The Brothers Karamazov* is not just fleeting, but even given as if by-the-way. More important, the narrator would have us believe, are the articles that Ivan has published since under the by-line "Eyewitness," and the most recent of them in particular. As the narrator explains:

> The incident was rather curious. Having already graduated from the university, and while preparing to go abroad on his two thousand rubles, Ivan Fyodorovich suddenly published in one of the big newspapers a strange article that attracted the attention even of non-experts, and above all on a subject apparently altogether foreign to him, since he had graduated in natural science. The article dealt with the issue of ecclesiastical courts, which was then being raised everywhere. (*Brothers* 16; *PSS* 14: 16)

If the issue was "then being raised everywhere," Ivan's decision to address it in his series of articles is hardly "sudden," nor is there anything "foreign" to the natural scientist in an article that, as the narrator adds, was at first favourably reviewed by "many churchmen" and then, "[s]uddenly, ... along with them, not only secularists but even atheists themselves" (16; *PSS* 14: 16). In short, where the narrator professes himself unable to make a coherent story out of these apparently disparate facts, we can: even the brief study of science together with the mention of atheism and an exposé style of journalistic writing add up to garden-variety nihilism. While we hear no more of Ivan's training in the natural sciences after that brief and understated mention, its apparent effects emerge most strikingly in Rakitin's conversations with Dmitri as the latter awaits trial.

Rakitin is one of the many reflections of Ivan's thought that make up much of the cast of characters in *The Brothers Karamazov*. Like the young Kolia Krasotkin and Ivan's half-brother Smerdiakov, Rakitin is not Ivan, but another version of who Ivan might be. When we first meet the thirteen-year-old Kolia, for example, and before he is fortunate enough to find a father-figure in Alyosha Karamazov, Kolia is an Ivan in the making. "I only respect mathematics and natural science" (*Brothers* 551; *PSS* 14: 497) Kolia says, before dismissing God as a "hypothesis" useful "for the sake of order" (553; *PSS* 14: 499); "I'm a socialist," he adds, "an incorrigible socialist" (554; *PSS* 14: 500). Rakitin is an older and seedier version of the same set of ideas. As a seminarian with socialist leanings who pays less than lip service to actual belief in God, Rakitin combines in himself the various audiences for Ivan's "strange" article; by the end of the novel Rakitin has turned to exposé-style journalism himself with the publication of an account of Dmitri's trial in the St. Petersburg tabloid *Rumors* (573; *PSS* 15: 14). Like Kolia, Rakitin is finally also a proponent of a certain kind of science, for example, when he insists that Alyosha in his innate sensuality is a "full-fledged Karamazov" after all: "so race and selection do mean something," he says (80; *PSS* 14: 74). Darwin in Rakitin's reading is not an advocate of Herzen's or Eliot's vision of radical contingency, but instead provides a natural law, a cause that necessarily produces a given result, as does his Claude Bernard.

When Alyosha visits Dmitri in prison, he is surprised at Dmitri's sudden question, "Who is this Carl Bernard?" "No, not Carl, wait," he adds, "I've got it wrong: Claude Bernard. What is it? Chemistry or something?" (*Brothers* 588; *PSS* 15: 28). It's of course physiology, not chemistry, but Dmitri has evidently grasped the basic idea. In Michael Katz's summary, Bernard "believed in the absolute determinism of natural science; in his words: 'the conditions of a phenomenon once known and fulfilled, the phenomenon must occur'" (Katz 22), and Rakitin seems to have explained as much. According to Dmitri, Rakitin plans to write an article "with a tendency: 'It was impossible for him not to kill, he was a victim of his environment'" (588; *PSS* 15: 28). As Dmitri attempts to explain:

Imagine: it's all there in the nerves, in the head, there are these nerves in the brain (devil take them!) ... there are little sorts of tails, these nerves have little tails, well, and when they start trembling there ... that is, you see, I look at something with my eyes, like this, and they start trembling, these little tails ... and when they tremble, an image appears, not at once, but in a moment, it takes a second, and then a certain moment appears, as

> it were, that is, not a moment – devil take the moment – but an image, that
> is, an object or an event, well, devil take it – and that's why I contemplate,
> and then think … because of the little tails, and not at all because I have a
> soul or am sort of image and likeness. (589; *PSS* 15: 28)

If Dmitri's confused presentation emphasizes the helplessness that the phrase "victim of his environment" also implies, for Rakitin as for the real Claude Bernard, the facts of material determinism deprive the ordinary man of free will only in order to lend the scientist a God-like power.

"It's magnificent, Alyosha, this science! The new man will come, I quite understand that," Dmitri says, although he can't help but feel "sorry for God": "Chemistry, brother, chemistry! Move over a little, Your Reverence, there's no help for it, chemistry's coming!" (*Brothers* 589; *PSS* 15: 28–9). Rakitin himself doesn't share Dmitri's misgivings. As Dmitri tells Alyosha: "'But,' I asked, 'how will man be after that? Without God and the future life? It means everything is permitted now, one can do anything?' 'Didn't you know?' he said. And he laughed. 'Everything is permitted to the intelligent man'" (589; *PSS* 15: 29). As Rakitin in Dmitri's rendering amplifies the bare mention of Ivan's scientific inclinations, so he makes clear what for Dostoevsky is the most fundamental, most frightening, and also most contradictory aspect of nihilist/positivist science: its claims to "extraordinariness."

The paradox of material determinism as nihilist/positivist science conceived it is that it doesn't apply to everyone equally: while the average individual is indeed at the mercy of the "laws of nature," the "intelligent man" is their master. Certainly, the "extraordinariness" that permits or even encourages murder is also a matter of man-made law, especially as Raskolnikov formulates it in *Crime and Punishment*. Raskolnikov, yet another writer, has apparently published an article on the topic, again, not that we are given to see it. In the words of the police investigator Porfiry Petrovich:

> The whole point is that in his article all people are somehow divided into
> the 'ordinary' and the 'extraordinary.' The ordinary must live in obedience
> and have no right to transgress the law, because they are, after all, ordi-
> nary. While the extraordinary have the right to commit all sorts of crimes
> and in various ways to transgress the law, because in point of fact they are
> extraordinary (*Crime* 259; *PSS* 6: 199).

The "law" that Porfiry Petrovich has in mind pertains to the legal system and social convention more generally, and Raskolnikov in his own restatement of his argument lists law-givers only in a legal sense: "the Lycurguses, the Solons, the Muhammeds, the Napoleons, and so forth" (*Crime* 260; *PSS* 6: 200).[3] In the age of Buckle and Quetelet, however, man-made and "natural" or scientific laws overlap, and Raskolnikov is quick to quantify his theory of transgression with the certainties of "simple arithmetic" (65; *PSS* 6: 54). In Raskolnikov's article, as he recalls it, his historical examples also give way to once again a recognizably Darwinian theory of evolution.

While Raskolnikov doesn't exactly disagree with Porfiry Petrovich's summary of his argument, he feels a need to restate it anyway, at first with a degree of hesitation. "The only difference," he begins:

> is that I do not at all insist that extraordinary people absolutely must and are duty bound at all times to do all sorts of excesses, as you say.... I merely suggested that an "extraordinary" man has the right ... that is, not an official right, but his own right, to allow his conscience to ... step over certain obstacles, and then only in the event that the fulfillment of his idea – sometimes perhaps salutary for the whole of mankind – calls for it. (*Crime* 259; *PSS* 6: 199)

As Raskolnikov warms to his argument, though, his language shifts from "right" to "nature," as in "by their very nature," "again by nature," and "according to the law of nature" (260; *PSS* 6: 200), until Porfiry Petrovich finally pushes him to explain himself. Just as Rakitin contemplates the effect of "race and selection," so Raskolnikov accounts for the "remarkably" small number of people born with the capacity "of saying anything *new*" with recourse to Darwin. As he explains:

> An enormous mass of people, of material, exists in the world only so that finally, through some effort, some as yet mysterious process, through some interbreeding of stocks and races, with great strain it may finally bring into the world, let's say, at least one somewhat independent man in a thousand. Perhaps one in ten thousand is born with a broader independence (I'm speaking approximately, graphically). With a still broader independence – one in a hundred thousand. Men of genius – one in millions; and great geniuses, the fulfillers of mankind – perhaps after the elapsing of many thousands of people on earth. In short, I have not looked into the retort where all this takes place. But there certainly is and must be a definite law; it can be no accident. (263; *PSS* 6: 202)

The same connection between "extraordinariness" and the "laws of nature" is made much more evident in Rakitin's visit to Dmitri in *The*

Brothers Karamazov. Even in Dostoevsky's last and longest novel, however, this all-important connection is expressed only indirectly.

In Dmitri's confused rendering, after all, it may be Claude or perhaps Carl, while Ivan's role as the real promulgator of the Bernardian belief in the scientist as "an inventor of phenomena, a real foreman of creation" has been deliberately obscured: it is not just that our not entirely reliable narrator passes so quickly over the issue of Ivan's actual training in the sciences, but that Ivan's claims to "extraordinariness," like Raskolnikov's article, are never presented quite in his own words. When Dmitri quotes Rakitin to the effect that "everything is permitted," for example, he quotes Rakitin quoting Ivan for the second time at secondhand.[4] Rakitin, like the reader, hears the line first from Ivan's cousin Miusov, and he then repeats it to Alyosha, quoting Miusov apparently quoting Ivan: "And did you hear his stupid theory just now: 'If there is no immortality of the soul, then there is no virtue, and therefore everything is permitted'" (*Brothers* 82; *PSS* 14: 76); only then does he repeat the phrase to Dmitri, although now as his own idea. If we ever hear Ivan speak the words attributed to him, it is only because we understand Ivan's devil as a hallucination and so in a more literal and less symbolic sense an aspect of Ivan himself. Even Ivan's devil, though, purports to quote someone else, not Ivan, but "a most charming and dear young Russian gentleman: a thinker and a great lover of literature and other fine things" (648; *PSS* 15: 83).

It's a transparent ruse, especially as the devil goes on to describe his young Russian gentleman as "the author of a promising poem entitled 'The Grand Inquisitor.'" The devil nonetheless adds another layer of confusion when he summarizes what he claims are the thoughts of his young thinker with reference not to the poem of "The Grand Inquisitor" that we already know, but to another that we don't: "Geological Cataclysm." "Remember that?" the devil asks, as he embarks on a summary of his young thinker's thoughts. In his paraphrase:

> Once mankind has renounced God, one and all (and I believe that this period, analogous to the geological periods, will come), then the entire old world view will fall of itself.... Man will be exalted with the spirit of divine, titanic pride, and the man-god will appear. Man, his will and his science no longer limited, conquering nature every hour, will thereby every hour experience such lofty delight as will replace for him all his former hopes of heavenly delight. (*Brothers* 648–9; *PSS* 15: 83)

The question that remains, the devil explains, is what to do in the meantime. When and if the era of unlimited science arrives, "then everything will be resolved and mankind will finally be settled." Until then, the young man decides, "anyone who already knows the truth is permitted to settle things for himself, absolutely as he wishes, on the new principles. In this sense, 'everything is permitted' to him" (649; *PSS* 15: 83–4), including, we understand, murder – in *The Brothers Karamazov* as in *Crime and Punishment*.

As we work through paraphrase and quotes within quotes to reconstruct the scientific underpinnings of "extraordinariness" in Raskolnikov as in Ivan Karamazov, Dostoevky's often ostentatious indirection serves at least two functions, both related to reading. That reading is perhaps more essentially ours, as Dostoevsky's echoes and repetitions engage his reader in a process of piecing together clues that reflects on a higher level the detecting that marks these admittedly unconventional murder mysteries. The reading is also, however, Dostoevsky's. As Robert Belknap has beautifully shown, Dostoevsky is remarkable not just for the sheer range of his reading, but for his astonishing ability to create in his writing clusters of associations that both indicate that reading and raise it to another level. As the quotes within quotes would suggest, his heroes' words really aren't their own. They are instead windows onto other texts.

In the case of "extraordinariness," those other texts include most obviously Balzac's *Père Goriot* (1835). As Leonid Grossman argues, it is Bianchon's thought experiment in *Père Goriot* that lies behind Raskolnikov's real murder for the greater good in *Crime and Punishment*; Dostoevskyan "extraordinariness" also looks to the example of Balzac's "exceptional" arch-criminal Vautrin. Vautrin is a man of near superhuman physical power, as evidenced in his astonishing self-discipline at the moment of his arrest:

> [T]he agents drew their pistols. [Vautrin] realized the danger ... and suddenly proved himself supremely endowed with human strength. Terrible and majestic sight! The phenomenon to be read in his face can only be compared to that of a boiler full of steam ... which a drop of cold water dissolves.... A murmur of reluctant admiration ran through the room as they saw how swiftly the lava and flames erupted and then withdrew into this human volcano. (Balzac 183)

His crimes, like Raskolnikov's, also invite comparison with Napoleon. Again like Raskolnikov, although without the benefit of Raskolnikov's education, Vautrin is finally enabled in his life of crime by his intuitive grasp of the social sciences that Balzac's novels helped invent. While Vautrin, like Lydgate, is a little too early for a Comte or Quetelet, still he credits his power in large part to what looks to twenty-first-century eyes like an anthropological stance. As he explains to Rastignac:

> Paris, you see, is like some forest in the New World, where a score of savage tribes run about like Illinois or Huron Indians, living off the spoils of their hunt in different social circles. You are hunting for millions.... There are many sorts of hunting; some hunt dowries, others hunt for profit-taking on their shares, some go fishing for consciences, others sell their subscribers all trussed up. (101)

"Extraordinariness" in Dostoevsky famously also refers to Rakhmetov in *What Is to Be Done?*, the one-time aristocrat and now full-time revolutionary who inspires Vera Pavlovna and her friends with his asceticism, superhuman strength, and full confidence in the victory of socialism: "Yes," Chernyshevsky's narrator tell us, "he was an extraordinary man, a specimen of a very rare breed" (Chernyshevsky, *What* 292). The "harder" natural sciences that Rakitin dabbles in gain more prominence in Chernyshevsky's vision of "extraordinariness," although Rakhmetov himself studies the natural sciences only briefly before leaving the actual practice of science to his marginally more ordinary disciples. "Extraordinariness" in an age of positivism finally finds a particularly striking embodiment in one of the great villains of the nineteenth-century novel: the stupendous Count Fosco in Wilkie Collins's *The Woman in White*.

Collins published *The Woman in White* as he wrote it, in installments that appeared starting in November 1859 in both Charles Dickens's *All the Year Round* and the American *Harper's Weekly*. The success was immediate and enormous, in Russia as in the West; in Russia, as Alexander Druzhinin wrote, "The Woman in White was one of the most widely read novels in all of 1861," "purchased and gulped down with more greed than Dickens's *Expectations* or *Framley Parsonage*" (Druzhinin 408). A large part of that success derived from the invention of Fosco. Like Dostoevsky, although more exceptionally among less Francophile British writers, Collins was a reader of Balzac and of French true-crime literature more broadly, and Fosco again owes something to Vautrin as well as to the Gothic villains of an earlier generation. That said, and as Collins's first readers immediately recognized, Fosco with his immense weight and his light tread, his flamboyant waistcoats and collection of tame animals, including a cockatoo, two canary birds, and a whole family of white mice, was also an instant type in his own person.

Fosco is in fact only the second although far more significant villain in a highly involved plot. The first villain, Sir Percival Glyde, marries

Laura Fairlie solely for her money and, when that money is not immediately forthcoming, hatches a diabolical plot at the instigation of his friend Fosco. This plot hinges on Lady Glyde's uncanny resemblance to the "Woman in White," Anne Catherick, a weak-minded young woman whom Sir Percival has already had institutionalized once. Now, at Fosco's urging, Sir Percival purports to send Anne back to the asylum, but instead sends his wife in her place. Meanwhile, as Fosco knows, Anne is already suffering from a heart complaint, and when she dies under the guise of Lady Glyde, Sir Percival inherits Laura's fortune. Luckily for Laura, however, she is loved by her erstwhile drawing master, Walter Hartright, who conspires with Laura's valiant half-sister Marian Halcombe to support Laura after her escape from the insane asylum, drive Sir Percival to his death, and force Count Fosco to France where he meets his end as a one-time member of an Italian revolutionary organization turned (presumably government) spy.

The very complication of the plot speaks to Fosco's imposing power and a ruthlessness cast in Napoleonic terms; as the ever-observant Marian notes in her diary, "He is a most remarkable likeness, on a large scale, of the Great Napoleon" (Collins, *Woman* 196). Again not coincidentally, Fosco is also a serious, if amateur, scientist. In the compilation of first-hand testimonies that makes up the quasi-legalistic structure of *The Woman in White*, Fosco speaks in his own voice only twice. Both times, however, he takes the opportunity to boast of his scientific credentials. When Fosco invades Marian's diary with his terrifying "[POSTSCRIPT BY A SINCERE FRIEND]" (307), he refers only briefly to his "vast knowledge of chemistry" and "luminous experience of the more subtle resources which medical and magnetic science have placed at the disposal of mankind" (308). In the written confession of the workings of the crime that Walter forces from him towards the very end, he explains at more length:

> The best years of my life have been passed in the ardent study of medical and chemical science. Chemistry, especially, has always had irresistible attractions for me, from the enormous, the illimitable power which the knowledge of it confers. Chemists, I assert it emphatically, might sway, if they pleased, the destinies of humanity. Let me explain this before I go further.
>
> Mind, they say, rules the world. But what rules the mind? The body. The body (follow me closely here) lies at the mercy of the most omnipotent of all mortal potentates – the Chemist. (560)

In our own age of psychotropic medication, we are familiar with the "vulgar" belief in thought or even personality as a by-product of chemical (im)balance. In its "magnetic" form, however, Fosco's material determinism well exceeds our own. Despite its association with the occult and the staged effects of the séance, "magnetism" or, as it is more often called, mesmerism, didn't allow for any kind of spirituality at all. Its claims of a "universal fluid" instead accounted for any and all phenomena in the flattest of material terms.

In *The Brothers Karamazov* Dostoevsky demonstrates the same move in terms of a false religiosity that conceives of a spiritual dimension only ever in material form. For Father Zosima's great rival Father Ferapont, holiness is a matter of the proper observance of fasts, and devils can be caught by slamming doors shut on their tails; at apparently the other end of the spectrum, the biological father Fyodor Pavlovich is mired in the same material world in his inability to imagine hell other than in the physical terms of life as we know it here on earth. If devils are to drag him down with hooks, he asks, "Where do they get them? What are they made of? Iron? Where do they forge them? Have they got some kind of factory down there?" (*Brothers* 24; *PSS* 14: 23). Mesmerism even at its height was widely regarded as a pseudo-science, a kind of wishful thinking rather than a real account of the natural world. If not entirely mainstream, still its appeal was exactly to materialists of the Ferapont- or Fyodor Pavlovich-ian sort, Balzac among them.

As Göran Blix argues, Balzac's fascination with the possibilities of a universal fluid derived from "the fluid's imbrication in matter, its close kinship with electricity, magnetic force, heat, light, and energy, all physical properties, and Balzac's instinctive hesitation before things that lacked positivity" (Blix 267). Like the fictional Fosco, the real Balzac also combined mesmerism with chemistry, repeating over and over his belief that "feeling condenses chemically into a fluid, perhaps one similar to that of electricity" (Blix 267). In his own contribution to the literature of mesmerism, *The Blithedale Romance*, Nathaniel Hawthorne exposes the sleight of hand that mesmerism performs when Miles Coverdale rejects Westervelt's show as offering "a delusive show of spirituality," while "yet really imbued throughout with a cold and dead materialism." Westervelt speaks of a coming era that "would link soul to soul," Coverdale says, but describes the means by which this goal would be achieved "in a strange, philosophical guise, with terms of art, as if it were a matter of chemical discovery." "[N]or would it have surprised me," Coverdale adds, "had he pretended to hold up a portion of his universally pervasive fluid, as he affirmed it to be, in a glass phial" (Hawthorne, *Blithedale* 185).

If Fosco's aim is an omnipotence achieved by entirely material means, like his less grandiose counterparts in the real world, he is nonetheless undone by the very "fluid" that is also his source of power. Although the practitioners of mesmerism were most often male, their subjects tended to be female, in the famous real-life case of Dr. John Elliotson and the O'Key sisters in the late 1830s, for example, as in the fictional pair of Westervelt and Priscilla in *The Blithedale Romance*, and the mesmeric bid for power more often than not only recast the male control of female bodies in pseudo-scientific terms. As Alison Winter details in *Mesmerized: Powers of Mind in Victorian Britain* (1998), those female and often working-class bodies sometimes also turned the tables, most spectacularly in the real case of the O'Keys. As the O'Keys extended their mesmeric repertoire to include different and far more strong-willed mesmeric personalities, the diagnosis of other patients, and even the wielding of mesmeric power themselves, Winter writes, "the interactions between operator and subject [became] necessarily more extended" (Winter 73) and even reversed; by the time a mesmerized Elizabeth O'Key was dictating to Elliotson how and when she would emerge from her trance, it began to seem that "the experimental subjects and the researchers had even changed places" (78). The clairvoyance that was popularly seen as a side effect also offered a means of shifting the power dynamic, one that Marian in *The Woman in White* uses to turn Fosco's mesmerism against him.

Fosco's science in all its forms assumes a strict separation between subject and object, when in the case of the redoubtable Marian those borders prove more than a little porous. Marian herself acknowledges the increasing control over both her mind and her body that Fosco finally realizes in his frightening postscript. As she plots her and Laura's escape, she is afraid even to sit near him: "His eyes seemed to reach my inmost soul," Marian writes, while "[h]is voice trembled along every nerve in my body, and turned me hot and cold alternately" (Collins, *Woman* 261–2). Fortunately for Marian and her sister, Fosco's practice of mesmerism unintentionally unleashes Marian's own clairvoyant powers in her prophetic dream of Walter's return. Marian also produces a finally fatal effect on Fosco when the attraction to her that he calls "the one weak place in [his] heart" leads to the discovery of "the one weak place in [his] scheme" (569).[5] Fosco's defeat all on its own suggests Collins's own investment in a less one-way and more "collective" kind of science, one where interactions between would-be operator and subject become not just more extended, but more truly interactive. This commitment becomes most clear, however, when we juxtapose the collapse of Fosco's "extraordinary" schemes in *The Woman in White* with Ezra Jennings's triumph in *The Moonstone*.

Like *The Woman in White*, *The Moonstone* offers not just a famously convoluted plot, but also a famously convoluted narration, as the mystery of Rachel Verinder's stolen diamond is told in thirteen parts by eleven different narrators, each of whom relates only as much of the plot as s/he witnessed first-hand. Only by novel's end does it become clear that the theft was perpetrated by Rachel's two suitors acting as an impromptu tag-team. One suitor, Godfrey Ablewhite, turns out to have been interested only in Rachel's fortune, as his subsequent attempts to raise money on her diamond reveal. The other, Franklin Blake, not only loves Rachel, but also removes the diamond from her room under the influence of a dose of opium that he doesn't know he has taken.

This highly involved mystery is solved in a "bold experiment" planned and executed by the marginalized figure of Ezra Jennings. Labouring under the burden of his own unfairly but irredeemably sullied reputation, Jennings can only find work as an assistant to a country doctor, and his scientific work "addressed to the members of my profession – a book on the intricate and delicate subject of the brain and the nervous system" (Collins, *Moonstone* 382) will never see the light of day. Jennings's vaguely Eastern origins and appearance, not to mention his addiction to opium, also associate him with the Indian diamond and with an Indian mysticism apparently at odds with good English science. Indeed, when Jennings first proposes awakening Blake's latent memory, the lawyer Mr. Bruff sees nothing but "a piece of trickery, akin to the trickery of mesmerism, clairvoyance, and the like" (410). Jennings insists, however, that what he offers is real nineteenth-century British science: "Science sanctions my proposal, fanciful as it may seem" (398), he tells Blake, before handing him extracts from the works of two real figures in British medicine, Elliotson again as well as Dr. William Benjamin Carpenter.[6]

In his preface to the novel, Collins appropriates Jennings's claim by emphasizing the empirical underpinnings of "the physiological experiment which occupies a prominent place in the closing scenes of *The Moonstone*" (Collins, *Moonstone* xxiii). It's not just empirical underpinnings at stake here, however. What the experiment shows is that while Blake objectively stole the diamond, subjectively he didn't. His responsibility for the theft is exactly like Ivan Karamazov's responsibility for the death of his father, only in reverse – where Blake committed the crime and yet didn't, Ivan didn't commit the crime and yet did. The science that affords such a multilayered view of crime and criminal intentions is necessarily one of a "softer," more unwieldy sort, lacking the certainties of Fosco's chemical, medical, and "magnetic" will to power. Our recognition of scientific possibilities beyond the "extraordinary" in *The Moonstone* as in *The Brothers Karamazov* is exactly what Schur has in mind when she asks that we "restore to the science of Dostoevsky's times some of its intellectual range and complexity" (Kaladiouk 419–20). It also suggests a different approach to the genre of detective fiction.

Despite all the detecting that Collins and even Dostoevsky offer, literary historians have tended to position both just a little off to the side of a genre that in its twentieth- and twenty-first-century incarnations stretches to accommodate a remarkably wide variety of writers, everyone from Agatha Christie and Raymond Chandler to Boris Akunin and Daria Dontsova. At issue is a tendency to limit the genre in its "classic" phase to the relatively narrow terms of Ryan's "deductive logic and forensic analysis." Just as nineteenth-century science was never a simple matter of "scientific method" and "systematic thinking" alone, however, so detective fiction was always a more complicated affair, even in the admittedly idealized world of Sherlock Holmes. For Umberto Eco the marker of the genre is not and never was a particular way of knowing, one that involves "superior powers of observation, vast scientific and human knowledge, and, above all, the use of scientific method" (Ryan 29). For Eco, the genre that he describes as of "all model plots," "the most metaphysical and philosophical" (Eco, *Postscript* 53) poses instead the entirely Dostoevskyan question of if and how we know anything at all.

The Science of Detecting

As Eco argues, for all the evident importance of the concurrent rise of forensic science, we define nineteenth-century detective fiction in terms of systematic thinking, deductive logic, and forensic analysis only by dint of a highly selective reading. Certain authors and texts are excluded from the start, notably among them Charles Dickens's *Bleak House*. *Bleak House* offers at least in part the mystery of the murder of Mr. Tulkinghorn, and it comes equipped with a professional detective in the form of Inspector Buckle; it also offers a proto–Sherlock Holmesian episode when the young Dr. Woodcourt briefly turns his hand to detecting. "And so your husband is a brickmaker?" Woodcourt asks:

"How do you know that, sir?" asks the woman, astonished.

"Why, I suppose so, from the color of the clay upon your bag and on your dress. And I know brickmakers go about working at piecework in different places. And I am sorry to say I have known them cruel to their wives, too." (Dickens 617)

For most readers it would nonetheless seem that *Bleak House* is too "novelistic" (read: high) for a genre most often defined in terms of formulae.[7] Readers looking for science only of a certain sort are also bound to be disappointed, and not just because Dickens includes a less-than-realistic episode of spontaneous combustion. More essentially, as Lawrence Frank argues, the science that runs through *Bleak House* is marked by its refusal to see meaning as anything other than radically contingent at best.

Jarndyce and Jarndyce, as the absurdly oxymoronic name of the case itself would suggest, is ultimately a matter of enormous bundles of papers that mean nothing at all, an apocalypse that produces no revelation. This problem of meaning also remains unresolved in the question of Esther's identity, as Esther finds a sense of self through her own near-death and the death of her recently discovered mother only to repress that self once more in the novel's final lines. As Frank notes, Esther's near-death experience produces a vision of meaninglessness as even a kind of refuge: "Dare I hint at that worse time," Esther recalls, "when, strung together somewhere in great black space, there was a flaming necklace, or ring, or starry circle of some kind, of which *I* was one of the beads! And when my only prayer was to be taken off the rest, and when it was such inexplicable agony and misery to be part of the dreadful thing?" (Dickens 480; L. Frank 98). For readers determined to read nineteenth-century detective fiction in terms of scientific objectivity, order, and resolution "once and for all," *Bleak House* obviously won't do. *The Moonstone* as the work of Dickens's friend, protégé, and eventually even relative by marriage fares only a little better with such readers, and for much the same reasons.

Historians of the genre most often quote T.S. Eliot's assessment of *The Moonstone* as "the first and greatest of English detective novels" only to refute his claim. While the general consensus acknowledges Collins's contribution in his creation of Sergeant Cuff, Jacques Barzun and Wendell Hertig Taylor express an equally widely held view when they insist: "*Pace* T. S. Eliot, this marvelous book is not 'the greatest English detective story.' It is a good mystery with unforgettable characters and fine melodrama, but Sgt. Cuff (copied from life) is not conspicuously a detective, and the clues, though fairly laid out from the beginning, satisfy only an antiquarian interest in ratiocination" (Barzun 137–8). The same criticisms finally apply also to Dostoevsky, only more so, although with the caveat that Dostoevsky's less-than-canonical standing is most often attributed not to some kind of inadequacy, but to the simple fact of his Russian-ness.

Readers have long recognized that Dostoevsky often poses puzzles for his reader, puzzles that in *Crime and Punishment* and *The Brothers*

Karamazov even involve murder. They have also noted that Porfiry Petrovich in *Crime and Punishment*, like Sergeant Cuff in *The Moonstone*, clearly suggests the hero-detective starting with Dupin, through Sherlock Holmes and on into the twentieth century: "[I]t's the old material method," Razumikhin says, "Last year he ran down a case involving a murder where almost all the traces were lost!" (*Crime* 246; *PSS* 6: 189). If Dostoevsky is nonetheless most often *not* read in terms of a Western genre of detective fiction, it is in large part because, as A.I. Reitblat argues, Dostoevsky operated in a cultural context marked by a considerably less well-developed sense of personal property and of legal culture and entirely lacking in any tradition of private detection. Louise McReynolds argues more broadly that a natively Russian genre of the "criminal novel" [уголовный роман] was shaped by the Russian legal reforms of the 1860s and a Russian legal code that, in McReynolds's words, "left the jurors free to differentiate the sin from the sinner, as the psychiatrists did the crime from the criminal" (L. McReynolds 60). The result, both claim, was a particularly Russian tendency to focus less on whodunit than why, and Reitblat gives as examples the titles alone of a few works by the now forgotten A.A. Shkliarevsky: "Why Did He Kill Them?" ("Отчего он убил их?," 1872), "What Ruined Him?" ("Что погубило его?," 1877), and "What Prompted Murder?" ("Что побудило к убийству?," 1879). As the example of Collins would suggest, however, Dostoevsky's refusal to play by the rules of "ratiocination" cannot be attributed to his Russian-ness alone.

It is Poe who invents the term when he calls the Dupin stories "tales of ratiocination," and the word has usually been taken to express a belief in order that informs detective fiction above all in its "classic," or nineteenth-century phase. Carl Malmgren writes that what he calls mystery fiction "unfolds in a rational world grounded in laws of cause and effect" (Malmgren 14), and Michael Holquist makes the point still more strongly, arguing that Poe "is the Columbus who lays open the world of radical rationality," and his detective "the essential metaphor for order," "the instrument of pure logic, able to triumph because he alone in a world of credulous men holds to the Scholastic principle of *adequatio rei et intellectus*, the adequation of mind to things, the belief that the mind, given enough time, can understand everything" (Holquist 156–7). Evidently a post-Enlightenment phenomenon, "ratiocination" is equally an investment in a certain kind of science.

The relationship with science is already apparent when Dupin at the climactic moment in "Murders in the Rue Morgue" turns to Cuvier's "minute anatomical and generally descriptive account of the large fulvous Ourang-Outang of the East Indian Islands" (Poe 498). More strikingly, as a great many scholars post-Foucault argue, the detective novel after Poe emerges alongside a science of criminology that promises an objective approach to policing based on the collection and identification of the material attributes of criminals and the material traces of their crimes. As Ronald Thomas puts it regarding *The Moonstone*, Collins's innovation is not Cuff, but Ezra Jennings, as Collins's is the "first novel of any kind to demonstrate in a compelling way the emergence of the modern field of forensic science and its growing importance to the new science of criminology" (Thomas 67). Certainly, the movement of doctors into detecting is a significant historical moment, including not just Drs. Jennings and Woodcourt, but also Watson and Conan Doyle. As McReynolds notes, in Russia as early as 1869 L.E. Vladimirov defended at Kharkov University the first dissertation on forensic expertise, "On the Significance of the Physician-Experts in Criminal Jurisprudence'" (L. McReynolds 61). To cast the science that undergirds nineteenth-century detecting only in terms of a paradoxical combination of "radical rationality" and an equally radical empiricism, however, is not just to relegate Dostoevsky and Collins along with Dickens to the margins of an ever-shrinking genre. It is also to read even Poe and Conan Doyle in the narrowest of terms.

For the likes of Barzun and Taylor, it is exactly in terms of forensic evidence that Collins, like Dostoevsky, falls short. In *The Moonstone*, Rachel with her eyewitness testimony is at least objectively right, and the stain on Franklin Blake's nightgown can be explained, although only in the most roundabout of fashions. Grigory's eyewitness account of the open door in *The Brothers Karamazov*, on the other hand, apparent proof that Dmitri had been in his father's room, is simply wrong, while "material" evidence in *The Brothers Karamazov* as in *Crime and Punishment* is entirely lacking, in the case of the latter despite Porfiry Petrovich's unsubstantiated claim to have discovered "a little trace" (*Crime* 458; *PSS* 6: 350). If the science and detection in Dostoevsky and Collins are too unwieldy to fit easily into a "world of radical rationality," however, I would argue that the problem is our own attachment to the very idea of "ratiocination." Not only is nineteenth-century science a great deal more multifaceted than the standard use of that term would suggest, but so too is nineteenth-century detective fiction.

In *Victorian Detective Fiction and the Nature of Evidence* (2003), Lawrence Frank recasts our assumptions by reading Poe and Conan Doyle together with Dickens and all three as responding not to the rise of forensic science but to a Darwinian theory of evolution that posited a universe of pure contingency "governed by chance *and* necessity" (L. Frank 4). This combination, Frank claims, lends itself to a fundamentally and also disturbingly historical outlook: "If detective fiction in the nineteenth century responded to various historical disciplines," Frank writes, "it revealed that in excavating the past, the detective encounters an infinite regress without a point of origin: there is no firm grounding for any hypothesis" (26). This is not Rakitin's Darwin, but Darwin as George Eliot and also Gillian Beer read him, a Darwin who lends himself to what Frank, like Beer, calls a "Romantic materialism" (5). In Frank's argument, this far more unsettling sort of science is also at work in nineteenth-century detective fiction generally.

Pace Holquist, Poe's "Murders in the Rue Morgue" as a founding text in the genre of detective fiction is in fact not easily read as an exemplar of either "radical rationality" or radical empiricism. In the absence of the medical trappings of a Dr. Watson or Conan Doyle, Dupin makes a show of scientific method with his allusions not just to Cuvier, but to the science of stereotomy and Comte's "late nebular cosmogony" (Poe 479). Like Holmes, Dupin also claims the "superior powers of observation" that Ryan associates with the detective as "our ideal of the pure scientist." Frank nonetheless argues that Poe's reference to the "late nebular cosmogony" in fact leads us away from the scientific resolution of the world's problems "once and for all" to suggest instead "the existence of a contingent universe denied the trappings of divinity ... as ... [the detective] ... sets out to render the apparently unintelligible intelligible without recourse to supernatural explanations" (L. Frank 30). Like Nancy Harrowitz, we might also note Dupin's and his narrator-friend's insistence on sitting alone in the dark in "the rather fantastic gloom of our common temper" (Poe 476). I would finally add that Dupin's much-vaunted powers of observation have no reality for the reader, for the simple reason that we are given no opportunity to test them.

In "The Murders in the Rue Morgue" the key piece of evidence is the broken nail and tuft of orangutan hair that Dupin shows to the narrator and so the reader only at the very end of the story, just before he wraps the case up; in "The Purloined Letter" (1844), as Barbara Johnson notes, Poe deliberately withholds the description, both internal and external, of the very letter that we seek. The clues, to borrow Barzun and Taylor's criteria, are not remotely "fairly laid out from the beginning," and while we can't hold Poe to the rules of a genre that he didn't yet know he was creating, still we might wonder if he didn't intend "ratiocination" as a kind of hoax all along. After all, while his narrator-friend claims that "[t]here was not a particle of *charlatanerie* about Dupin" (Poe 478), Poe himself was well-known for his tricksy ways, even in Russia.

When Dostoevsky in 1861 introduced his own readers to an American writer still little known in Russia, he laid emphasis on what he presents as the very American materiality of Poe's particular brand of fantasy.[8] He notes in particular Poe's most famous hoax, the article published in *The New York Sun* in 1844 that falsely claimed that the real European balloonist Monck Mason had succeeded in flying across the Atlantic Ocean in a mere seventy-five hours. As Dostoevsky writes, "That description was done in such detail, so exactly, it was filled with such unexpected, incidental facts, it had such an appearance of reality, that everyone believed in that journey, needless to say, only for a few hours; then it of course turned out on inquiry that there had never been any journey and that Edgar Poe's story was a newspaper hoax" (*PSS* 19: 89). To Poe's own great pleasure, "The Facts in the Case of M. Valdemar" (1845) with its bizarre tale of mesmerism *"in articolo mortis"* enjoyed a similar degree of credulity, while his apparently non-fiction essay "Eureka" (1848), "on the Spiritual and Material Universe," has long suffered the opposite fate: readers to this day can't quite agree whether or not Poe really intended his "Prose Poem" to be believed.

Poe's predilection for hoaxes would suggest a sly joke at work when Dupin in the opening pages of "The Murders in the Rue Morgue" retraces the "course" of his narrator-friend's meditations by retracing their actual steps through the streets of a Paris that partakes of a very mixed reality. As Paul Woolf points out, while we can follow Dupin and his friend past some of the real landmarks that he names, some are misplaced and others are entirely imaginary. In Woolf's reading, the "interior private world of the mind" (Harrowitz 191) that Poe only pretends to put into direct correspondence with an outside world proves the real reality, and the same could also be said of "The Purloined Letter," where Dupin achieves his stated aim of a kind of mind-meld with a culprit who, in the case of Minister D—, turns out to be little more than a double of himself. In short, we might do better to follow Frank's lead and take Poe at his word only in a text that was obviously not meant to believed, the 1849 "Mellonta Tauta." "Mellonta Tauta" purports to be a letter from the year 2848 that presents science, including the writer's own, as a matter not of "ratiocination," but of pure guesswork. As for the "truth of Gravitation," our correspondent Pundita writes, "Newton owed it to Kepler," while "Kepler admitted that his three laws were *guessed at* – these three laws of all laws which led the great Inglitch mathematician to his principle, the basis of all physical principle – to go behind which we must enter the Kingdom of Metaphysics: Kepler guessed – that is to say *imagined*" (Poe 920). In contrast to Poe, Conan Doyle may well have intended that we believe what his detective says. Even Sherlock Holmes, though, offers a more complicated science of detecting than his popular reputation would suggest.

"Ratiocination" for Holmes as for Dupin is a two-part process where closely observed facts are immediately organized into a sometimes elaborate but apparently entirely rational theory that would explain them. While this method of building up evidence might more properly be called "inductive," Holmes himself tends to call it "deductive," as in, "observation shows me that you have been to the Wigmore Street Post-Office this morning, but deduction lets me know that when there you dispatched a telegram" (Doyle 56). As Lawrence Frank notes, however, Holmes, like Dupin, is in fact far from operating on the evidence of material traces alone. Just as Hutter argues that the opium-addicted Ezra Jennings is "the ultimate detective" in *The Moonstone* "precisely because he is able to see both the significance of the most trivial details and to allow his mind to wander past the boundaries of rational thought" (Hutter 191), so Frank associates Holmes's only somewhat less regular recourse to narcotics with "Romantic preoccupations with dreams and the mystery of human consciousness" (L. Frank 193). Frank also draws special attention to the gaps in Holmes's cases as in Watson's narratives, including the letter that Watson readily admits is missing from his account in "The Hound of the Baskervilles" (1902). These gaps are finally also fundamental to the group of scholars who banded together with Eco and Thomas A. Sebeok to produce *The Sign of Three: Dupin, Holmes, Peirce* (1983), although this collective approaches the problem from a different angle. Rather than re-reading Holmes's Darwinist underpinnings, Eco, Sebeok et al. focus on a method that they call not "deductive" or "inductive," but "abductive," a term borrowed from the nineteenth-century American philosopher Charles Sanders Peirce.

As Peirce explains, abduction "introduces a new idea": "Deduction proves that something *must* be; Induction shows that something *actually is* operative; Abduction merely suggests that something *may be*" (Harrowitz 181). As the Sebeoks point out, close observation of the "reddish mould" adhering to Watson's instep can tell Holmes that he was in the vicinity of the Wigmore Street Post-Office but not that he actually went in, let alone that he sent a telegram rather than met a friend or simply stopped by. Where Holmes's fans most often admire what they see as Holmes's commitment to empirically observed material evidence, "[w]hat makes Sherlock Holmes so successful at detection," the Sebeoks write, "is not that he never guesses but that he guesses so well" (Sebeok and Umiker-Sebeok 22), in fact, invariably correctly; as Holmes himself at one point admits, "we balance probabilities and choose the most likely. It is the scientific use of the imagination" (Doyle 106). We know that Holmes has once again imagined the right answer because Dr. Watson says so and because Watson's narrative function, as Eco argues, is to confirm "the correspondence between the Possible World imagined by the detective and the Real World" (Eco, *Limits* 160).

"Ratiocination" in this reading is not a form of "radical rationality," but instead, as Harrowitz writes, "an operative which can cut through various levels of reality, a creative reverie which transcends positivistic reason and assumptions" (Harrowitz 195). As it works with clues of another kind, so this "operative" also lends itself to more complicated answers. In his commentary to his own post-modern contribution to the genre, *The Name of the Rose* (1980), Eco argues that the order that detective novels offer stands not in simple opposition to chaos, but comes instead in various stripes. The labyrinth that is the monastery library in *The Name of the Rose*, Eco explains, is what he calls a "mannerist maze": in a "model of the trial-and-error process," "[t]here is only one exit, but you can get it wrong." The actual world as Eco's hero-detective comes to know it, however, is a labyrinth possessed of what Deleuze and Guattari call a "rhizome structure": "The rhizome is so constructed that every path can be connected with every other one. It has no center, no periphery, no exit, because it is potentially infinite." Unlike the monastery library, this greater world "can be structured but is never structured definitively" (Eco, *Postscript* 57–8). If this other kind of labyrinth is still a story of science, in its open-endedness it is emphatically not a science that would "solve the world's problems 'once and for all.'" It looks instead much more like science as Bruno Latour imagines it.

In *Pandora's Hope* (1999), Latour argues that the story that would explain the natural world is not a copy of that world along the lines of a realist painting: there is, he says, no *adequatio*. For Latour what science offers instead is a chain of transformations that "link us to an aligned, transformed, constructed world" (Latour 79), a point that Collins makes in his own way in *The Moonstone*. Before it turns out to be right, Franklin Blake's confusion of "Objective-Subjective and Subjective-Objective" is a running joke. "This question has two sides," Blake says, "An Objective side, and a Subjective side. Which are we to take?" (Collins, *Moonstone* 46–7), or: "Rachel's conduct is perfectly intelligible, if you will only do her the common justice to take the Objective view first, and the Subjective view next, and the Objective-Subjective view to wind up with" (180). Jennings's experiment, however, turns what Gabriel Betteredge calls Blake's "German-English gibberish" into solid and successful science. To put it in Latourian terms, once we move away from the model of a "mind-in-a-vat" and a reality "out there," ideas of objective and subjective no longer prove terribly useful; what matters instead is *"the quality of the chain in its entirety"* (Latour 69). As Franklin Blake learns, "[a]n essential property of this chain is that it must remain *reversible*…. If the chain is interrupted at any point, it ceases to transport truth – ceases, that is, to produce, to construct, to trace, and to conduct it" (Latour 69).

The Moonstone makes reference to other, less reversible and less interactive kinds of nineteenth-century science, for example, when Betteredge expresses his appreciation that Blake for all his faults has yet to dabble in dissection. "You see my young master, or my young mistress, poring over one of their spiders' insides with a magnifying glass," Betteredge says, "or you meet one of their frogs walking down stairs without his head; and when you wonder what this cruel nastiness means, you are told that it means a taste in my young master or my young mistress for natural history" (Collins, *Moonstone* 55). A Fosco-like chemical bid for power also hovers off-stage in the life story of Rachel's uncle, the Wicked Colonel who first brought the diamond to England. When not smoking opium or carousing in the slums of London, we are told, the Colonel "had dissipated the greater part of his fortune in his chemical investigations"; while he leaves the diamond to Rachel with ill intentions, his will also specifies that his remaining fortune be divided between the care and safe keeping of his animals and the endowment of "a professorship of experimental chemistry at a northern university" (41). When Ezra Jennings, in Mr. Bruff's legal phrase, "proves his case," however, it is because he practises a different kind of science, one that allows him first to reconstruct Dr. Candy's fragmentary memories, and then to show that Rachel's eyewitness testimony is not objectively wrong, but incomplete. It is a fundamentally Latourian approach where "only a tiny portion … rests on resemblance, on *adequatio*. All the others depend only on the conservation of traces that establish a reversible route" (Latour 61).

Jennings's evidently cutting-edge work "on the intricate and delicate subject of the brain and the nervous system'" would seem ideally suited to the broader shift that Latour advocates, one that begins when the "mind-in-vat" is "offer[ed]… a body again" and "the reassembled aggregate" is put "back into relation with a world that is no longer a spectacle at which we gaze but a lived, self-evident, and unreflexive extension of our ourselves" (Latour 8). As we will see in the next chapter, Dostoevsky, like Collins and also like Eliot, is similarly invested in the complicated intersubjective realities of minds and bodies. Still, when Dostoevsky tells a story of this sort of a more complicated and more contingent kind of science in so many words, the early glimmerings of a science of physiological psychology are only implied. The science that Dostoevsky calls by name is instead what in the 1860s and 1870s was a very new and still controversial discovery in mathematics: non-Euclidean geometry.

Dostoevsky and the New Math

At least since Descartes and all through the nineteenth century, including in Dostoevsky, a more conventional kind of mathematics has tended to serve as an image *not* of an embodied mind in the world, but exactly of Latour's "mind-in-a-vat." While Porfiry Petrovich is lightly ironic when he expresses the desire for "a piece of evidence that's something like two times two is four! Something like direct and indisputable proof!" (*Crime* 338; *PSS* 6: 261), the Underground Man loudly protests the mathematical claims of the "laws of nature" that would deprive him of his own free will. "Needless to say," he says mockingly, "all human actions will then be calculated according to these laws, mathematically, like a table of logarithms, up to 108,000, and entered into a calendar; or, better still, some well-meaning publications will appear, like the present-day encyclopedic dictionaries, in which everything will be so precisely calculated and designated that there will no longer be any actions or adventures in the world" (*Notes* 24; *PSS* 5: 113). As Dostoevsky knew well, however, even as Comte and Quetelet were inspiring a radically reductive mathematical quantification of "living life," more qualified and also more interesting mathematicians were engaged in other sorts of explorations.

A very different kind of mathematics is reflected late in *War and Peace* when Tolstoy makes a stunning turn to a "new branch of mathematics" that, "having attained to the art of dealing with the infinitely small," can now solve "other more complex problems of movement which used to appear insoluble" (L. Tolstoy 821, amended; L.N. Tolstoi, *PSS* 11: 266–7). The branch of mathematics that Tolstoy has in mind is calculus. Like Buckle, although, he claimed, independently of that effort, Tolstoy in *War and Peace* aspired to what he also calls "historical science" (822). Aside from his openly Buckle-esque ambitions, we might also note a distinctly positivist/nihilist tendency running through much of Tolstoy's thought, what Isaiah Berlin calls his "love of the concrete, the empirical, the verifiable" (10). In *Anna Karenina* (*Анна Каренина*, 1877), Levin's plan in the wake of the emancipation of the serfs to harness his workers' self-interest co-opts Vera Pavlovna's socialist sewing collective for entirely conservative ends, for example, while in *War and Peace* Tolstoy returns again and again to the project of scientific quantification. Just as "[f]orce (the quantity of motion) is the product of mass times velocity," Tolstoy writes, so "[i]n military action the force of an army is also a product of mass times something, some unknown x" (1034). Tolstoy himself identifies that "unknown x" with the "spirit of the army" (1034), an apparently intangible force that he strives to make scientifically quantifiable with the parenthetical and French gloss *"chaleur latente"* (790); his aim in accounting for military victories as in the movement of history more generally is always to find a law that can be expressed in mathematical terms. As the sheer length of *War and Peace* would suggest, however, Tolstoy is anything but reductive, and in his recourse to calculus he seeks a formula that would far exceed anything in Comte or Quetelet for suppleness and sophistication.

Tolstoy's grasp of the intricacies of calculus is a matter of some debate, as is its precise relevance to the largely literary work that is *War and Peace*.[9] His sources, however, have been well-delineated, including Laplace and possibly Helmholtz, as well as most certainly Tolstoy's friend and mentor in all things mathematical, Prince S.S. Urusov (1827–97), artillery officer, chess player, and author of a number of different works, including *Differential Equations* (*Дифференциальные и разностные уравнения*, 1863), *On the Integral Factor in Differential Equations* (*Об интегральном множителе разностных и дифференциальных уравнений*, 1865), *On Solving the Problem of the Knight* (*О решении проблемы коня,*1867), and *A Survey of the Campaigns of 1812 and 1813, Military-Mathematical Problems, and Concerning the Railroads* (*Обзор кампаний 1812 и 1813 годов, военно-математические зидичи и о железных дорогах*, 1868). The general outline of Tolstoy's ambitions would also seem clear. As Tolstoy explains at the very beginning of the third volume of *War and Peace*, the difficulty that we face in establishing a science of history is that "[f]or human reason, absolute continuity of movement is incomprehensible" (L. Tolstoy 821). Calculus has solved the problem of the continuity of motion in mathematics, he says, "by allowing for an infinitesimal quantity and the ascending progression from that up to one tenth, and by taking the sum of that geometrical progression" (821). Tolstoy argues that historians should now do the same. "Only by admitting an infinitesimal unit for observation," he writes, "a differential of history, that is, the uniform strivings of people – and attaining to the art of integrating them (taking the sums of these infinitesimal quantities) can we hope to comprehend the laws of history" (822). As Stephen Ahearn notes, while Tolstoy uses the older language of Leibnitz's infinitesimals, his proposal that historians find the sum of "infinitesimally small units" corresponds to what in the 1860s was the still very new idea of taking the limit of the Riemann sums.[10] However Tolstoy might have understood his "infinitesimals," though, "[t]his integration problem," Ahearn writes, "is quite difficult and Tolstoy gives no indication of how we might solve it" (634). In an echo of the problem in *Bleak House*, by the end of the novel any kind of solution at all also seems increasingly unlikely.

The second and last epilogue of Tolstoy's very long novel ends with a somewhat wistful comparison of history with astronomy, the latter still the scientific gold standard as Comte's first and most regular of all the sciences as quantified mathematically. As he starts the last chapter of his epilogue, Tolstoy explains once more that, just as Copernicus put an end to an older science of the planets, so:

> Ever since the first man said and proved that the number of births or crimes obeys mathematical laws, that certain geographic and politico-economic conditions determine one or another form of government, and that certain relations of the inhabitants to the land produce movements of peoples – ever since then the foundations on which history had been built were essentially destroyed. (L. Tolstoy 1213)

Even at the end of this very long book, however, Tolstoy doesn't quite have anything to put in their place. The penultimate paragraph of his last chapter begins, "As in the question of astronomy, then, so now in the question of history, all the difference in views is based on the recognition or non-recognition of an absolute unit serving as a measure of visible phenomena." That final chapter then ends with a last sentence that does anything but wrap the novel up: "In the first case, the need was to renounce the consciousness of a nonexistent immobility in space and recognize a movement we do not feel; in the present case, it is just as necessary to renounce a nonexistent freedom and recognize a dependence we do not feel" (1215). In this oddly uncompelling statement of a scientific law that Tolstoy after more than a thousand pages is still not quite ready to embrace, we hear the bitterness of a man who, in Berlin's famous assessment, "was by nature a fox, but believed in being a hedgehog" (Berlin 4). As Berlin writes, like the hedgehog, Tolstoy "longed for a universal explanatory principle" (36), when, like the fox, "[h]is genius lay in the perception of specific properties, the almost inexpressible individual quality in virtue of which the given object is uniquely different from all others" (36).[11] For Berlin, Dostoevsky was in contrast a hedgehog alone. I would argue instead, however, that Dostoevsky was simply much more comfortable with a truth that is both true and also multiple, as he was with the latest in nineteenth-century mathematics.

Neither is true for his hero Ivan, who introduces the topic of non-Euclidean geometry in *The Brothers Karamazov* only to reject it. In his call to rebellion in the lead-up to "The Grand Inquisitor," Ivan begins by assuring Alyosha that he accepts God "pure and simple." "But this," he adds, "needs to be noted":

if God exists and if he indeed created the earth, then, as we know perfectly well, he created it in accordance with Euclidean geometry, and he created human reason with a conception of only three dimensions of space. At the same time there were and are even now geometers and philosophers ... who doubt that the whole universe, or even more broadly, the whole of being, was created purely in accordance with Euclidean geometry; they even dare to dream that two parallel lines, which according to Euclid cannot possibly meet on earth, may perhaps meet somewhere in infinity. I, my dear, have come to the conclusion that if I cannot understand even that, then it is not for me to understand about God. I humbly confess that I do not have any ability to resolve such questions, I have a Euclidean mind, an earthly mind, and therefore it is not for us to resolve things that are not of this world. (*Brothers* 235; *PSS* 4: 214)

If Ivan's Euclidean mind is insufficient to understand God's purpose, he nonetheless plans to stick with what he knows. As he explains:

I have a childlike conviction that the sufferings will be healed and smoothed over, that the whole offensive comedy of human contradictions will disappear like a pitiful mirage, a vile concoction of man's Euclidean mind, feeble and puny as an atom, and that ultimately, at the world's finale, in the moment of eternal harmony, there will occur and be revealed something so precious that it will suffice for all hearts ... to justify everything that has happened with men – let this, let all of this come true and be revealed, but I do not accept it and do not want to accept it! Let the parallel lines even meet before my own eyes: I shall look and say, yes, they meet, and still I will not accept it. (235–6; *PSS* 14: 214–15)

In his work as a whole Dostoevsky's rejection of what Beer calls "that form of positivism which refused to acknowledge possibilities beyond the present and apparent world" (Beer, *Darwin's Plots* 142) has most often been read in religious terms. As scholars from E.I. Kiiko to Anna Schur and Vladimir Gubailovskii argue, however, Ivan's reference to non-Euclidean geometry also reflects Dostoevsky's early training and lifelong reading in the sciences.

While we have no evidence one way or the other, most scholars assume that Dostoevsky encountered non-Euclidean geometry in a review that appeared in the journal *Knowledge* (*Знание*) in 1876 written by the physicist and physiologist Hermann von Helmholtz. We know that Chernyshevsky did, because we have his response. While quick to acknowledge Helmholtz as "one of the greatest of naturalists," with the nihilist/positivist certainty that Ivan can't quite muster, Chernyshevsky dismisses non-Euclidean geometry itself as "childish waggery, not worthy of attention" (Kiiko 123). Although Helmholtz was widely read in Russia as in Europe, Dostoevsky might also have encountered George Henry Lewes's response to Helmholtz in *Problems of Life and Mind*, the first two volumes of which also appeared in Russian translation in *Knowledge* in 1876, complete with the appendix "On Imaginary Geometry and the Truth of Axioms."

From a scientific point of view, another world is not an option, and the challenge that "Imaginary Geometry" posed especially to nihilist/positivist thought was two-fold. At the very least the development of an entire geometry that dispensed with Euclid's fifth (or parallel) postulate threatened to split scientific rationality from empirical experience. What is still worse is the possibility, as science has since confirmed, that space as we think we know it, the kind where parallel lines don't meet, may not be what space is at all. As Peter Gaffney writes: "Not only does this mean the end of Newtonian universality (the claim that physical laws are applicable throughout time and space), challenging claims and assumptions based on the unity of science, but also it means the end of a mechanistic worldview in which matter passively fills out a set of determinate spatio-temporal relations" (17). A committed Deleuzian, Gaffney argues further that "a particular (historically specific) body of scientific thought has a reciprocal relationship with the object it determines, each one participating in the actualization of the other and simultaneously traversing a diversity of social, intellectual, and material processes" (3–4). Where Chernyshevsky can understand this sort of reciprocity only as a sort of joke or intellectual parlor game, a more serious sort of nineteenth-century science advocated exactly Franklin Blake's confusion of "Objective-Subjective and Subjective-Objective." It is also this possibility that Ivan refuses to entertain.

Not just despite but because of his increasingly desperate hold on a narrowly positivist/nihilist commitment to the strict separation of mind and material world, Ivan ends by losing both mind and matter. On the one hand, as Schur notes, Ivan claims an allegiance to empirical experience so extreme as to rule out any kind of understanding altogether: "I don't understand anything ... and I no longer want to understand anything," he says (*Brothers* 243; *PSS* 14: 222). As Schur argues, even in the context of contemporary positivism, Ivan's devotion to empiricism is so radical as to undermine that other pillar of positivist thought, rationality; indeed, Ivan speaks "as in a delirium." When Ivan refuses to accept even the evidence of his own eyes, however – "Let the parallel lines even meet before my own eyes: I shall look and say, yes, they meet, and still I will not accept it" – his problem is not a commitment to fact at the expense of reason, but a commitment to a certain kind of rationality at the expense of what he actually sees (236; *PSS* 14: 212). Dostoevsky's most immediate answer to this unhappy young nihilist operates in terms of the "living life" here and now that is, for Dostoevsky, also a belief in God: Alyosha kisses him. We then wait several hundred more pages before Ivan's devil gives a more explicitly scientific response.

While Ivan's devil may be a hallucination, as we will see in the next chapter, in the Lewesian terms laid out in *Problems of Life and Mind*, that doesn't mean that he isn't real. He is also quick to make recourse to mathematics. "[L]ike you, I myself suffer from the fantastic," the devil tells Ivan, "and that is why I love your earthly realism. Here you have it all outlined, here you have the formula, here you have geometry, and with us it's all indeterminate equations!" (*Brothers* 638; *PSS* 15: 73). Ivan would reject equations with multiple answers just as he rejects a space that he not only doesn't know but doesn't want to know. When the devil offers his own story about Paradise, Ivan can't believe that the atheist sentenced to walk a quadrillion kilometers has already reached the gates of heaven. "Arrived!" he asks, "But where did he get a billion years?" "You keep thinking about our present earth!" the devil says (644; *PSS* 15: 79). Like his father with his hooks, Ivan can't imagine a spiritual realm that exists other than in material terms; as the devil himself says in one of his best lines, "The other world and material proofs, la-di-da!" (636–7; *PSS* 15: 71). Ivan's limitation is also, however, a refusal to even consider the possibility that a material world might manifest itself in any other way than the one that he thinks he already knows.

In "On Imaginary Geometry" Lewes ends his only very qualified acceptance of non-Euclidean geometry by dismissing any formal axiomatic system that doesn't also partake of empirical proof. "Genii compressible into bottles, and expansible into giants, can be written about and pictured," he writes, "but they are not possible realities, which any fisherman may pull up in his net" (Lewes, *Problems* 2: 465). If God is exactly the genie that Dostoevsky pulls up in his net, that it's also faith doesn't mean that it's not science. As Kiiko points out, Ivan's rejection of God together with the latest in nineteenth-century mathematics suggests instead that religion and science, in certain non-Euclidean forms, might be compatible. Kiiko argues for the mutual implication of Dostoevsky's science, and his religion in his reading of two notes that Dostoevsky wrote to himself on 17 August 1880, shortly after completing work on the chapter "The Devil. Ivan Fyodorovich's Nightmare." The second note is especially striking:

> The real (created) world is finite, while the immaterial world is infinite. If parallel lines were to intersect, the law of this world would end.
>
> But they intersect in infinity, and infinity undoubtedly exists. For if infinity didn't exist, neither would finiteness, it would be meaningless. And if infinity exists, then there is a God and a world other than the real (created) world, one that is based on other laws.

In Kiiko's summary: "And so, the existence of God and of the 'other world' results from the recognition of the infinity of space, for which non-Euclidean laws, laws other than those for Earth, are true" (Kiiko 126). As Liza Knapp puts it, Dostoevsky finds in non-Euclidean geometry "a geometric embodiment of the yearning for infinity that he felt in his heart" (Knapp, *Inertia* 218). He also finds a fully scientific form of irresolution.

It is, after all, irresolution that Dostoevsky promises in *The Brothers Karamazov*, starting with the problem of his hero as he presents it in his prefatory "From the Author." Dark laughter runs all through Dostoevsky's professed concern that readers will find his Alexei Fyodorovich "a figure of an indefinite, indeterminate sort" (*Brothers* 3; *PSS* 14: 5). As he accounts for his apparently unconventional hero, Dostoevsky trips over himself to make ever more absurdly oxymoronic qualifications. "One thing, perhaps, is rather doubtless," he says, that his hero "is a strange man, even an odd one." If "rather doubtless" is already a confusing formulation, Dostoevsky then calls "odd" into question, explaining that an "odd man" is not only "'not always'" an exception, but that it even often happens that "it is precisely he, perhaps, who bears within himself the heart of the whole" (3; *PSS* 14: 5). Dostoevsky goes on to warn that his again very long novel is fundamentally incomplete, "even almost not a novel at all," he writes, "but just one moment from my hero's early youth" (3–4; *PSS* 14: 6). The problem is, Dostoevsky explains, that while he has just one biography, he has two novels, the second and more important as yet unwritten. "Thus my original difficulty becomes even more complicated," he writes: "for if I, that is, the biographer himself, think that even one novel may, perhaps, be unwarranted for such a humble and indefinite hero, then how will it look if I appear with two; and what can explain such presumption on my part?" The again absurdly repetitive answer that Dostoevsky gives offers fair warning for the novel that will ensue: "Being at a loss to resolve these questions," he says, "I am resolved to leave them without any resolution" (4; *PSS* 15: 6). This irresolution is not without its legal, moral, and even scientific rigour. There is nonetheless very little that we can say about the crime at the centre of the detective novel that is *The Brothers Karamazov* without a great deal of qualification and often complicated contextualization.

If by murder we mean the act of clocking Fyodor Pavlovich in the head with a paperweight, then it would seem to be Smerdiakov who is guilty. If we add in the murderous intentions that encouraged and even enabled Smerdiakov, however, the question of responsibility begins to look a great deal murkier. Ivan himself tries to confess to the murder while still sick with the brain fever that also gives rise to his hallucination of the devil. As I have already noted, in terms of the Lewesian physiological psychology that we will encounter in the next chapter, to argue that it is madness doesn't mean that it's not real. Ivan himself has also already complicated our understanding of what an "act" might entail with his memory of doing nothing on the last night of his father's life:

> Remembering this night long afterwards, Ivan Fyodorovich recalled with particular disgust how he suddenly would get up from the sofa and quietly, as though terribly afraid of being seen, open the door go out to the head of the stairs, and listen to Fyodor Pavlovich moving around below, wandering through the downstairs rooms.... All his life afterwards he referred to this "action" as "loathsome," and all his life, deep in himself, in the inmost part of his soul, he considered it the basest action of his whole life. (*Brothers* 276; *PSS* 14: 251)

It is finally Dmitri who is convicted of the crime, although in the twelfth and last "book" of the novel, pointedly titled "A Judicial Error."

Dostoevsky's narrator takes his title from the defence attorney's concluding remarks: "Gentlemen of the jury," Fetyukovich warns, "beware of a judicial error" (*Brothers* 740; *PSS* 15: 166). As the reader by this time knows, the defence attorney Fetyukovich is correct in his claim that the chain of entirely circumstantial evidence against Dmitri is "fantastic," not an air-tight legal case but something more like "a novelistic suggestion" (731; *PSS* 15: 157). As the reader also knows, however, the "error" is also Fetyukovich's own assertion that in any case "[s]uch a murder is not a murder" (747; *PSS* 15: 172); like Rakitin, Fetyukovich claims the amorality of positivist/nihilist material determinism: "It was impossible for him not to kill, he was a victim of his environment" (588; *PSS* 15: 28). The novel ends with the possibility of Dmitri's escape to America still in the offing. That said, Dmitri's again qualified "hymn" to the suffering that he is not entirely ready to endure reflects not least a recognition of his own role in the affair. Like Ivan, Dmitri may not have wielded the murder weapon, but he wanted his father dead, and his wild words to that effect gave Smerdiakov both the cover and the opening that he needed.

Just as the development of non-Euclidean geometry derives from our recognition that "the four postulates of absolute geometry simply do not pin down the meanings of the terms 'point' and 'line'" and "that there is room for *different extensions* of the notions" (Hofstadter 222), so any statement of guilt or responsibility is contingent on what we take the words "guilt" and "responsibility" to mean. Ivan's reference to non-Euclidean geometry, brief as it is, accordingly encapsulates the problem of the novel as a whole: truth is not less true for being multiple, for being rooted in the particularities of real life, and even, as Gary Saul Morson argues, for taking in "the entire *field* of possibilities." For Dostoevsky, Morson writes, "to understand an event is to grasp not only what incident happened to happen, but also what others might have happened…. [T]o understand a person is not only to know what life he did lead, but also what others he might have led: we are all capable of leading more lives than one" (Morson 481). Ivan's mathematical struggle with other sorts of spaces speaks to this field of possibilities. It also reflects Dostoevsky's fundamental belief in another world that is as real as our own.

This other world is not real as mesmerism would have it, in the exact material terms of the world that we claim to already know. The devil arrives to mock Ivan's nihilist/positivist insistence on a very restricted kind of tangibility on every level, from the story of another time and space to his dream that he might be incarnated "so that it's final, irrevocable, in some fat, two-hundred-and-fifty-pound merchant's wife" (*Brothers* 639; *PSS* 15: 74). As the devil knows, for Ivan, as for Rakitin, for his father and for Father Ferapont, more matter and more flesh is the only measure of more reality. As the devil himself as a manifestation not least of Ivan's impending physical collapse also illustrates, neither is this other world real as spirit alone, an abstraction cut off from the everyday experience of our own bodies. It is instead Karamazovian "sensuality" in its ideal form, not the father's crude sexuality with his spitting and the bobbing of his Adam's apple, but the Feuerbachian embodiment of spirit in and by "living life" that is as instinctive to Ivan as it is to his brothers. "Sticky spring leaves, the blue sky – I love them, that's all!" Ivan admits. "Do you understand anything of this blather, Alyoshka, or not?" "I understand it all too well," Alyosha answers, "to want to love with your insides, your guts – you said it beautifully, and I'm terribly glad that you want so much to live….. I think that everyone should love life before everything else in the whole world." "Half of your work is done and acquired, Ivan," he adds. All that remains is "[r]esurrecting your dead, who may never have died" (230–1; *PSS* 14: 210).

In *The Brothers Karamazov* the dead would include Ivan's father, who at that point in the novel is already murdered in thought and in word, if not yet in deed. The dead are also the real bodies in the world in all their particularity, including Ivan's own, that nihilist/positivist science leaves just outside its purview. In *The Brothers Karamazov* Dostoevsky only barely tells the story of an Ezra Jennings–like science that would reflect the multiplicity and open-endedness of living bodies in the world. While his brief allusion to non-Euclidean geometry implies it, still we unfold that meaning only by way of a good deal of literary as well as scientific context, from Dickens, Collins, and Balzac to Helmholtz and Lewes. It is Dostoevsky's special genius that he packs his reading and his thought into dense clusters of allusions, including not just Ivan's Euclidean mind but, as Robert Belknap argues, the Rousseauesque implications of Switzerland or, as we will discuss in chapter five, Svidrigailov's use of "Schiller" as a term of abuse. Dostoevsky makes the point of another sort of science most strikingly, however, not when he names that science by name but when he embodies and also enacts it in the shape and workings of his prose. As Ponge promises, form is the real frontier, in the actual spaces that non-Euclidean geometry delineates. As we will see in the next chapter, for Dostoevsky an alternative science is less a matter of what he says in his plots than of plot itself as a literary device.

3 Bodies and Plots

An excessive reliance on plot has long served the literary establishment as a marker of "bad" literature. As Peter Brooks summarizes the critical consensus in his *Reading for The Plot* (1984):

> "Reading for the plot," we learned somewhere in the course of our schooling, is a low form of activity…. Plot has been disdained as the element of narrative that least sets off and defines high art – indeed, plot is that which especially characterizes popular mass-consumption literature: plot is why we read *Jaws*, but not Henry James. (Brooks 4)

Brooks attributes this critical tendency to "the advent of Modernism" and an early twentieth-century "suspicion toward plot," one "engendered perhaps by an overelaboration of and overdependence on plots in the nineteenth century" (7). While nineteenth-century novels, including Dostoevsky's, are indeed known for their large casts of characters and intersecting storylines, still the critical disdain that Brooks describes is not a twentieth- or twenty-first-century phenomenon. It is instead already at work in the nineteenth century itself.

In Russia, Tolstoy at one end of the literary-political spectrum and Chernyshevsky at the other offer especially striking examples of nineteenth-century writers who disparage the workings of plot even while availing themselves of its possibilities. Despite Tolstoy's memorable way with plot, most notably in the "peace" parts of *War and Peace* as in the multiple storylines of *Anna Karenina*, still his eventual decision to leave off fiction writing altogether reiterates in the most tangible of terms the Chernyshevskyan claim that serious writers make recourse to plot only as a concession to an unfortunately vulgar and still insufficiently educated reading public. As the narrator of *What Is to Be Done?* explains with maximum clarity:

Yes, the first pages of my story reveal that I have a very poor opinion of my public. I employed the conventional ruse of a novelist: I began my tale with some striking scenes taken from the middle or the end, and I shrouded them with mystery. You, the public, are kind, very kind indeed, and therefore undiscriminating and slow-witted.... So I was obliged to bait my hook with striking scenes. (Chernyshevsky, *What* 47)

If Dostoevsky with his tangled murder mysteries, tales of intrigue and illicit relationships, and often shocking scandal scenes in contrast embraces plot, it is not because he was less attuned than either Tolstoy or Chernyshevsky to the "overelaboration of and overdependence on plots" that by the mid-nineteenth century was already a cliché.[1] While Dostoevsky most definitely sought some degree of mass consumption, it is also unlikely that a writer of such philosophical and formally inventive bent made literary choices based entirely on what he thought would sell. This chapter will argue that plot and plot-tiness serve Dostoevsky instead as yet another means of expressing his commitment to the open-endedness and multiplicity of living bodies in the world, now as reflected in another newly emerging science: physiological psychology.

Where Dostoevsky, as we have seen, combined his "sense of the clustering mystery of a material universe" with his faith in another set of possibilities altogether, George Henry Lewes and Hermann von Helmholtz, as the authors of *Problems of Life and Mind* and *On the Sensations of Tone as a Physiological Basis for the Theory of Music* (*Die Lehre von den Tonempfindungen als physiologische Grundlage für die Theorie der Musik*, 1863) respectively, restricted themselves to the former alone, and they put the problem of non-Euclidean geometry squarely in the context of ever-present issues of mind and body and mind and material world. It may well be, as Lewes suggests in "On Imaginary Geometry" and as mathematicians today would agree, that "the homaloidal Space with which Geometry deals is in fact a curved Space, the curvature becoming sensible when very distant points are taken" (Lewes, *Problems* 2: 462). In Lewes's argument, however, what we call "objective" material reality is in any case only real to the extent that a given mind and body can know it. Lewes's larger claim in the main body of the work to which "On Imaginary Geometry" is only appended, *Problems of Life and Mind*, is that "the external world exists, and *among* the modes of its existence is the one we perceive." For what he calls "other forms of Sentience (if there are such) than our own" (1: 168), reality takes on a very different shape that is no less real. In *On the Sensations of Tone* Helmholtz makes the same point when he argues that musical taste is a matter not just of cultural norms but of an anatomy that he is careful to cast as equally relative. "In the whole of this research," he writes, "we have dealt solely with natural phenomena, which present themselves mechanically, without any choice, to all living beings whose ears are constructed on the same anatomical plan as our own" (Helmholtz 234). In nineteenth-century Europe the interrelationship of mind and material world that Lewes made especially famous found significant literary expression in Lewes's own fiction as well as in the writing of his common-law wife George Eliot. It also shaped yet another new literary genre: the novel of sensation.

The novel of sensation as it emerged in the wake of Collins's *The Woman in White* has often been cast in Chernyshevskyan terms as an appeal to a body that can't be trusted, a "baiting of the hook" that would bypass rational thought to speak to readers' bodies alone. As D.A. Miller describes it, the genre offered "one of the first instances of modern literature to address itself primarily to the sympathetic nervous system, where it grounds its characteristic adrenalin effects: accelerated heart rate and respiration, increased blood pressure, the pallor resulting from vasoconstriction, and so on" (D. Miller 146). These "characteristic adrenalin effects" are usually attributed to a particularly lurid set of plot elements, including, as Richard Fantina writes, "the themes of inheritance, bigamy, poisoning, drug abuse, and adultery, and ... frequent employment of the *deus ex machina* and other startlingly improbable coincidences" (Fantina 23). In this reading, the aim of the novel of sensation is to sell "bad" literature to a largely female and otherwise intellectually impoverished readership whose bodies lack the mental discipline to resist its corporeal effects. We do at least Collins a disservice, however, if we understand "sensation" either as science or as literature in "vulgar" terms alone.

After all, as the last chapter has already argued, it is only his villain Fosco who believes that minds are ruled by bodies, while Collins himself advocates the mutual implication of mind and material world that marks the work of the real Lewes as it does that of the fictional Ezra Jennings. It is also not the case that "sensation" and its attendant plot-tiness are tools wielded only by writers of a "low" and accordingly commercially oriented sort. George Eliot's reputation for high-mindedness has long obscured a leaning toward sensation that scholars most often discern in her less well-known works, including *Romola* (1862) and the very odd and apparently un-Eliot-like *The Lifted Veil*. As Kate Flint argues in her sharp reading of Eliot at her most canonical, however, even in *Middlemarch* "whether sensationalism occurs in the shocks that a plot may deliver, or in the metaphors through which wordless mental agony is conveyed to the reader, the body, as the locus of sensation, remains a constant and necessary given" (Flint, "Materiality" 78). The same is also true for Dostoevsky.

Dostoevsky presents the often improbable events that make up his plots as both reflecting and also shaping his characters' conjoined physical and mental states, most strikingly in the many illnesses that afflict his characters, including Ivan's brain fever in *The Brothers Karamazov*, Myshkin's "idiocy" in *The Idiot*, and the "morbid" state that Raskolnikov experiences for much of *Crime and Punishment*. They have the same effect on the combined bodies and minds of his readers. Dostoevsky deliberately sets up his machines not just to encourage an ideal reading on the part of an abstract or implied reader but also to produce a visceral effect in flesh-and-blood readers, the latter not at the expense of the former but as part of a Lewesian and also Latourian project to restore minds and bodies to one another. We can only measure that effect in subjective terms. Still, anecdotal evidence would suggest that he still achieves that end.

A 2001 survey of readers in Cheliabinsk includes among the "associations" that readers make with Dostoevsky not just physical items (axes) and topoi (St. Petersburg), but also "sensations" [ощущения] and "states" [состояния] such as "pain," "sickness," "nerves," and "hysteria" (Zagidullina 533–4). My own informal survey finds a similar reaction among the readers that I know best, my students. One student memorably compared reading the "strains" or "lacerations" chapters in *The Brothers Karamazov* to the sound of fingernails scraping across a blackboard; after a course that featured *Notes from Underground*, one evaluation read: "No more Dostoevsky…. Painful book to read. Cringing." It is exactly this corporeal response that Dostoevsky aimed to achieve, and if the result is often what the Cheliabinsk survey describes as "discomfort" [дискомфорт] (527), it is not because Dostoevsky aspired to unpleasant sensations alone. Despite the evident success of his appeal to readers' bodies, it was also not the intent of this most philosophical of nineteenth-century writers to bypass cognition altogether. Dostoevsky's reliance on an often extreme level of plot-tiness instead reflects his belief that we truly think only when we *also* feel. As the Underground Man says, "where there is no love, there is no reason" (*Notes* 95; *PSS* 5: 157): for Dostoevsky as for Lewes and Latour, thought, real thought, and even real morality can only be grounded in the material realities of the world.

Death and Dismemberment

Dostoevsky is equally well known for the extravagance of his plots and for novels that enjoy little in the way of plot at all. Either way he draws our attention to the functioning of plot as literary device. This theoretical level is perhaps most evident in the case of the latter, for example, in *Notes from Underground* where, as Chloë Kitzinger argues, even as the "events" of the novel are largely little more than the unnamed narrator's rants, still those rants tell the story of the narrator's repeated attempts to construct a scenario where his role is that of the hero. If larger amounts of content tend to obscure that point, still Dostoevsky's plots, even at their most involved, always operate on that same meta-level. Dostoevsky's work in this other vein has most often been associated with his reading of French literature – the "boulevard" novel as exemplified by Eugène Sue's *The Mysteries of Paris* (*Les Mystères de Paris*, 1843), for example, or the French genre of "true" crime. In their high degree of self-awareness, however, Dostoevsky's plot-tier plots also indicate his extensive reading in contemporary British literature.

Dostoevsky's complicated recourse to the genre of the detective novel, for example, reflects not just a British-inflected plot, but a British-inflected commentary on plot. As Brooks argues, the detective story is "the narrative of narratives ... a laying-bare of the structure of all narrative in that it dramatizes the role of *sjužet* and *fabula* and the nature of their relation." The story at hand, he writes, is a repetition and also re-ordering of the story that was, a re-telling that only ends with the "apprehension of the original plotmaker, the criminal" (Brooks 25). Dostoevsky's earliest efforts post-Siberia are also explicit in their acknowledgment of another form newly pioneered by the British: the novel of sensation. When Mizinchikov in *The Village of Stepanchikovo* (*Село Степанчиково и его обитатели*, 1859) reveals to the narrator his intention to abduct Tatiana Ivanovna, he explains: "like Gretna Green, you understand" (Dostoevsky, *Village* 146; *PSS* 2: 538). Dostoevsky's narrator does understand: like Mizinchikov, he is evidently well versed in the British trope of elopements to the real Scottish village just over the border from England where a looser marriage law applied. By the same token, the child heroine whose tragic death ends the tale of unsanctioned love and a rich man's evil machinations that is *The Insulted and Injured* (*Униженные и оскорбленные*, 1861) is called "Nelli" by her mother, the late Jenny Smith. Certainly, Dostoevsky was reading Dickens.[2] He was also, however, reading a set of British writers who have historically enjoyed much less in the way of critical favour, including not just Collins but the novelists of "sensation" who emerged in his wake.

Yelizaveta Akhmatova's (1820–1904) exceptionally long-lived journal *Collected Foreign Novels, Novellas and Stories in Russian Translation* (*Собрание иностранных романов, повестей и рассказов в переводе на русский язык*, 1856–85) led the way in introducing the Russian reading public to a set of British writers who included not just Collins but also Mary Elizabeth Braddon and Ellen, or Mrs. Henry Wood (1814–87), and whose plots featured not just elopements to Gretna Green, but also adultery, crime, and even murder. What we might consider more first-tier journals were quick to follow. Throughout the 1860s and 1870s, Katkov in the *Russian Herald*, for example, not only published Collins but also introduced Russian readers to lesser-known figures like Charles Reade (1814–84) and Dinah Maria Craik (1826–87). The dramatic twists and turns of plot that mark Dostoevsky's mature works, including murder, suicide, and the violation of young girls, as well as the "scandal scenes" that Bakhtin describes – the money cast into the fire at Nastasya Filippovna's name-day party in *The Idiot*, or the catastrophe that is the fête in *Demons* – reflect this new kind of writing. They also stand in stark contrast to the style preferred by Dostoevsky's two great rivals in the Russian literary scene: Tolstoy and Chernyshevsky.

Plot is not Chernyshevsky's strong suit, and Dostoevsky regularly pokes fun at his wooden characters and stilted scenarios. In *Notes from Underground* the narrator's obsessively planned bumping of shoulders with a man in the street parodies Lopukhov's similar display of manliness, while in *Crime and Punishment* Chernyshevsky's Crystal Palace reappears as a seedy bar; in *Demons* the same "progressive" picnic that in *What Is to Be Done?* is all merriment and laughter ends in utter unseemliness. While Tolstoy, with his mastery of complicated plot and characterization, might seem Chernyshevsky's polar opposite, it is his tragedy that Tolstoy combined a remarkable gift for plot and characterization with a great deal of anxiety at the physical effects of art, especially when it comes to female bodies. Tolstoy often presents the non-verbal arts as conduits for an unthinking and unbridled hedonism, in the adulterous effects of the opera in *War and Peace*, for example, or of Anna's portrait in *Anna Karenina*, or the possibly adulterous and certainly murderous effects of the Beethoven duet in *The Kreutzer Sonata* (*Крейцерова соната*, 1889). He was also increasingly ill at ease with the effects of his own literary efforts, even across the course of just one admittedly very long work: *War and Peace*.

The famously strange mix of more traditionally novelistic elements with an array of non-fiction writing that makes up *War and Peace* is in fact more than a little unbalanced, as for the first half of the novel Tolstoy largely confines himself to "bait[ing his] hook with striking scenes." It is only roughly halfway through that the narrator intervenes to lay bare the abstract idea that his plot, as it turns out, only illustrates that history is made up not of the actions of "so-called great men" (L. Tolstoy 606), but of the "unconscious, swarm-like life of mankind" (605). As the telling that is Tolstoy's essayistic writing then increasingly replaces the showing that is his story, that telling itself also becomes ever more abstract. By the entirely plot-less Part Two of the Epilogue, it's not just that Tolstoy has moved from Pierre and Natasha to Copernicus and a science of history, but that he has stripped even his images of their pseudo-material trappings. We reach a kind of end point, for example, when he reveals the counterfeit rubles that first appeared as an element of plot and of historical fact (609) as nothing more than an image of the "paper currency" that matters most, the kind that historians have exchanged "for the pure gold of actual understanding" (1187). Whatever other impulses drive Tolstoy's stripping away of his own art – and we might note that a particularly virulent strain of self-criticism runs all through his art, as it does his life – in this instance they also overlap with his nihilist tendencies.

Where Tolstoy, despite his best efforts, never quite succeeded in sacrificing his gift for plot and character on the altar of an idealized and entirely rational communication, the entire enterprise comes much more easily to Chernyshevsky. As lacking as *What Is to Be Done?* might be in literary qualities, from Chernyshevsky's unabashedly nihilist/positivist point of view, the issue is not that he can't produce art, but that he won't. Chernyshevsky's proto-revolutionary Rakhmetov can protest as much as he likes that his weakness for cigars only demonstrates that he, too, is "not an abstract idea, but a human being, one who longs to live life" (Chernyshevsky, *What* 290), but the narrator readily acknowledges that his "extraordinary" man, too, is a "type," albeit "a specimen of a very rare breed" (292), as are the only slightly less remarkable Lopukhov and Kirsanov. In the same way, Chernyshevsky's rejection of the plot that he lifted from George Sand in the first place is quite openly an appeal to the "perspicacious" reader, one who has trained him- or herself to do without the sugar-coating of "striking scenes" and "embellishments" (48). As the narrator explains, "I possess not one bit of artistic talent. I even lack full command of the language" (48), and if his starkly didactic intentions make for artificial characters and a cumbersome plot, his goal is to do without "such vulgarity" (47) altogether.

It is certainly true that a nineteenth-century Russian writer, and especially one already in prison for his seditious views, wouldn't succeed in publishing a call to revolution without the fig leaf of fictionality. Still, in an argument that both Brooks and Tolstoy would recognize, what Chernyshevsky emphasizes instead are the "poor instincts" (Chernyshevsky, *What* 47) of a still inadequate reading public. "Good, strong, honest, capable people," he writes:

> – you have only just begun to appear among us; already there's a fair number of you and it's growing all the time. If you were my entire audience, there'd be no need for me to write. If you did not yet exist, it would be impossible for me to write. But you're not yet my entire audience, although some of you are numbered among my readers. Therefore, it's still necessary and already possible for me to write. (49)

As his critics have long noted, for all Chernyshevsky's claims to the primacy of matter, his entire utopian project rests on the assumption that the more "perspicacious" among his readers will find the intellectual means to resist their own "poor instincts" and re-educate them in others. What Chernyshevsky aspires to in his readers he also models in his characters.

For Chernyshevsky as an avowed materialist, the "real" reality is not literature, but the material world that is the so-called object of science, including not just the bodies that respond to similar chemical stimuli in the same way but also the environmental conditions that dictate particular forms of behaviour. Chernyshevsky accordingly lays great emphasis not just on the physical dimensions of his heroes – Lopukhov's broad shoulders, for example, or Vera Pavlovna's ample bosom – but also on the actual functioning of their bodies. When a troubled Lopukhov takes two morphine pills to help him sleep, for example, he finds that his "spiritual travail was roughly equivalent in strength (according to Lopukhov's materialist viewpoint) to four cups of strong coffee" (Chernyshevsky, *What* 252), while Vera Pavlovna and her second husband repeatedly interrupt their discussions of physiology with what would seem discreet reference to actual sexual encounters.[3] Even as he emphasizes the materiality of his fictional characters and real readers, however, it is exactly the body that slips out, as Chernyshevsky's repeated recourse to theoretical intervention directly contradicts his own premises.

Even as a materialist Chernyshevsky can never leave matter well enough alone, but instead tries to jump-start its evolution with ever more explanation, not just in terms of his repeated and heavy-handed calls to the "perspicacious reader," but in the very premise of a novel that purports to end in a utopian future that takes place two years after the novel's actual publication date. In exactly the same fashion, the bodies that he portrays *en route* to the Crystal Palace seldom manage to make their way there all on their own. They are instead regularly disciplined by the rigours of theory, especially in the case of female bodies with their often unruly emotions, or Rakhmetov with his bed of nails. As Chloë Kitzinger argues, it is largely because Chernyshevsky eschews literariness along with what she calls the "messiness and contingency" of actual human beings that *What Is to Be Done?* was so successful "at producing positive models of human behavior that 1860s Russian readers could actively use." In comparison, Kitzinger writes, *Crime and Punishment* attracted far fewer adherents, and understandably so; in her words, "one work is designed for practical dismemberment; the other, for aesthetic and mimetic coherence" (Kitzinger 158). For all its gestures at transparency, however, Chernyshevsky's repeated and strenuous recourse to "types" only pretends to engage with actual living matter while really offering pseudo-scientific abstraction. If Chernyshevsky's lapses in logic are only a little tragic and more often ludicrous, still the comparison with Tolstoy is illustrative. In fact, most materialisms go that way.

Leaving aside the Marxist variant that, as Jane Bennett writes, swaps actual matter for intangible economic relationships, the "scientific" or "vulgar" materialism of the "you are what you eat" variety is particularly striking in this regard. The "vulgar" drive to equate living organisms with the inanimate matter that they ingest and excrete offers remarkably easy answers to what might seem complicated questions, from the apparently inevitable outcomes of British imperialism to the workings of plot and characterization in *What Is to Be Done?* These answers also tend to suit the almost invariably left-wing politics of the so-called scientists, so much so, in fact, that it would seem obvious that the theoretical conclusions of "vulgar" materialism precede or even act entirely in the absence of any empirical evidence. As their would-be material monism reverts always to a set of theoretical assumptions, this quasi-scientific over-determination has the curious effect of emptying "vulgar" materialism of any real matter at all. As Dostoevsky with his call for "living life" was quick to see, it also tends to empty living organisms of any actual life. Like "life-expectancy tables" and "statistical data," references to "nitrogeneous creatine" and even Moleschott's "potato blood" only pretend to engage with actual living matter while really offering pseudo-scientific abstraction, a retreat from the very life that nihilist science purports to explain. This reversal is still more striking in Bernard's *Introduction to the Study of Experimental Medicine.*

While Claude Bernard was a far more serious and scientifically grounded advocate of material determinism than Vogt, Büchner, or even Moleschott, still his attempt to reduce all of life to the law-like effects of quantifiable material conditions produces a similarly "vulgar" result. Although physiology, as Bernard explains, is "the science whose object is to study the phenomena of living beings and to *determine* the material conditions in which they appear," Bernard was controversial in his own day for the practice that he delicately termed "dissociation," and in the *Introduction* he tackles the issue head-on (Bernard 66). To give his most famous quote once more, "[i]f a comparison were required to express my idea of the science of life," he writes, "I should say that it is a superb and dazzlingly lighted hall which may be reached only by passing through a long and ghastly kitchen" (15). As he explains, over the course of his work the physiologist might "detach living tissues, and … place them in conditions where we can better study their characteristics." "We occasionally isolate an organ by using anesthetics to destroy the reactions of its general group," he adds, or "reach the same result by cutting the nerves leading to a part, but preserving the blood vessels" (88–9). As he notes in another section, "This is what we observe when we place a small animal under an air pump; its lungs are obstructed by the gases liberated in the blood" (120). Unfortunately for the small animals under the physiologist's care, "[t]o extend his knowledge," Bernard writes, "he has had to increase the power of his organs by means of special appliances; at the same time he has equipped himself with various instruments enabling him to penetrate inside of bodies, to dissociate them and to study their hidden parts" (5).[4]

"Living life," as it turns out, can do much worse than it does in *What Is to Be Done?* Just as his plot is an obvious artifice and his materially determined heroes little more than do-gooding automata, so Chernyshevsky makes a point of mangling his characters' bodies on the rack of theory. In Bernard's hands, however, once-living bodies are actually cut up into pieces, a still more dire fate that finds its most direct literary representation in Émile Zola's (1840–1902) *Thérèse Raquin* (1867).[5] In an essay that he published in Russia in 1879, "The Experimental Novel" ("Le Roman Expérimental"), Zola makes the extraordinary claim that his project is exactly Bernard's, so much so, he writes, that "[i]t will often be but necessary to replace the word 'doctor' by the word 'novelist,' to make my meaning clear and to give it the rigidity of scientific truth" (Zola 1–2). Just as Bernard's experimental medicine studies "the phenomena of living beings" in order "to *determine* the material conditions in which they appear" (Bernard 66), so "experimental," or "naturalistic novelists," Zola explains, "are the determinists who experimentally try to determine the condition of the phenomena, without departing in our investigations from the laws of nature" (Zola 30). We might note, and many readers have, that Zola's insistence on equating real and fictional "experiments" reflects a fundamental misunderstanding of what science does. What is nonetheless strikingly Bernardian in *Thérèse Raquin* is the way that a "naturalistic" attempt to represent life results in a representation of death instead.

If the bodies in *What Is to Be Done?* are surprisingly abstract, in *Thérèse Raquin* they are actually dead, not just in the case of Camille, whose murder is anticipated and then recalled over and over again, but especially in the long, detailed, and gruesome descriptions of the heaps of dead flesh that Laurent encounters in his repeated visits to the morgue. Just as Chernyshevsky holds out the promise of the physical pleasures of the Crystal Palace in Vera Pavlovna's fourth dream, so Bernard assures us that the end of science is a "superb and dazzlingly lighted hall." Unfortunately, Zola's representation of a world of matter alone, including the living matter of bodies, never makes it out of the "ghastly kitchen." In *Dying to Know: Scientific Epistemology and Narrative in Victorian England* (2002), his own exploration of the conjoined workings of nineteenth-century literature and science, George Levine argues that what he calls a "desire for death" is a "central aspect of modern civilization"; "[o]r perhaps it might be said," he adds, "that modern scientific epistemology is a perfect medium through which the desire for death might get expressed" (Levine 2). In the literary terms that also make up Levine's case, not just Zola's morgue, but Tolstoy's and Chernyshevsky's distrust of plot reflects the same belief that we access truth not through, but despite "living life." If "vulgar" materialism is quite literally a dead end, however, "modern scientific epistemology" doesn't have to go that way.

While Levine makes a partial exception for George Eliot, he presents only William James as operating in entirely different terms. "Against the almost masochistic objectivist model," he writes, "William James wrote philosophy like a novel, finding the story of self-sacrificial surrender to an inhuman world both hard to take and untrue" (Levine 31); in a suggestive phrase, he also describes James as "unabashedly narrative" (41). While recognizing James as a model, however, and exactly with Eliot in mind, I would contrast "the almost masochistic objectivist model" not with narrative in general, but with a particular kind of narrative, one marked by a high degree of plot. Like Beer, I would also argue that plot and plot-tiness of this sort are far more widespread in nineteenth-century thought than Levine seems to recognize.

In *Darwin's Plots* Beer argues Darwin's debt to *Bleak House*, "with its apparently unruly superfluity of material gradually and retrospectively revealing itself as order, its superfecundity of instance serving as an argument which can reveal itself only *through* instance and relations" (Beer, *Darwin's Plots* 6). Darwin, she writes, was freed by his reading of Dickens, "whose style insists upon the recalcitrance of objects – their way of mimicking the human order without yielding their own 'haecceitas'" (56). In Beer's account, Dickens's turn towards rather than away from plot entails not just an insistent materiality, but a particular attention to the reader. "Dickens and other novelists such as Elizabeth Gaskell," Beer writes, "even sought physically to affect their reader: we are to laugh and weep as we read: rictus and wetness. We are to be physically disarranged by the reading experience" (41). "Though this may seem a far cry from Darwin's emphasis on substantiation," she adds, "there is the identical drive towards confirming experience by appeal to the physical and the material, changing language into physical process. We see another form of it in the 'sensation' novel" (41). This "appeal to the physical and the material" was not just a means of selling more books, although it certainly seems to have succeeded in that. It also reflects yet another emerging science, physiology not as Moleschott or Bernard imagined it, but the physiological psychology made famous by a figure that we have already encountered: George Henry Lewes.

The Science of Sensation

Well before he joined the non-Euclidean fray, Lewes was a major player in the emerging fields of physiology and physiological psychology, and his admittedly qualified openness to the possibilities of "On Imaginary Geometry" was an extension of his work on the material underpinnings of perception. Lewes was certainly not without his positivist/nihilist side. Like Bernard, he was an active and vocal vivisectionist. As a one-time advocate of Comte and like the "vulgar" materialists, Lewes was also often associated with left-wing politics, especially in Russia, indeed so much so that he makes a fleeting appearance in *Crime and Punishment* when Lebeziatnikov recommends that Sonia read his 1859 *The Physiology of Common Life*. As this cameo appearance would indicate, Dostoevsky was well aware of Lewes's reputation in nihilist circles. Even so, a highly positive if unsigned review of *The Physiology of Common Life* that appeared in Dostoevsky's journal *Time* in 1861 suggests that Dostoevsky himself read Lewes in a different light.

This review begins by praising Lewes's genuine difficulty. Homegrown physiologists of the cruder Chernyshevskyan sort would do best to steer clear of the English, the anonymous author warns: "All these Leweses, Buckles, Millses, Darwins are a most dangerous people," he writes. Where the Russians apparently prefer to rush ahead without waiting for hard evidence, he explains, the English are "so cold-blooded, cautious and mistrustful, such sceptics and people so little likely to be carried away, that you won't get very far with them" ("Fiziologiia obydennoi zhizni" 51). Indeed, Lewes's actual practice as a scientist provided a level of complexity that eluded the more "vulgar" materialists in Russia as elsewhere. As Rick Rylance argues in *Victorian Psychology and British Culture, 1850–1880* (2000), Lewes's still cutting-edge science was also a consequence of his "versatile, polymathic, innovative" approach (Rylance 251). In an age when science was already becoming a professional pursuit, Lewes was self-taught and unaffiliated with any institution. He also refused to specialize, as he not only actively fostered the career of his common-law wife George Eliot, but was himself the author of a much-noted biography of Goethe, histories of philosophy and of theatre, a great deal of literary criticism, and even a few early novels. Lewes was finally utterly unlike his nihilist/positivist contemporaries in his insistence that science offers no final word.

For Lewes, as Rylance explains, science as an area of inquiry was defined by its "provisionality, openness, revisability, and collective endeavor," and these formal principles are enshrined in the story that his science tells (Rylance 301). In a newly emerging field of psychology with the nihilists at one extreme and the Kantians at the other, Lewes's great contribution was his belief, as Richard Menke writes, that "physiology and psychology, nerves and neuroses, are best understood as, respectively, the objective and subjective presentations of what are in fact the same phenomena" (Menke 623). Turned outwards, Lewes's argument is the very modern idea that perception and what we might call the material world mutually inflect one another. While Lewes made his ultimate claim for the interaction of mind and material world in *Problems of Life and Mind*, what Menke calls his "dual-aspect monism" is also already on view in his early novel *Ranthorpe* (1847), published in Russian as *A Poet's Life* (*Жизнь поэта*) in 1859.

If Lewes was never the novelist that he helped George Eliot to become, still the hodgepodge of theme and plot that makes up *Ranthorpe* accurately conveys the pan-European literary context that was also Dostoevsky's. The Russian title, *A Poet's Life*, emphasizes the extent to which Percy Ranthorpe's story offers a re-telling of Balzac's *Lost Illusions* (*Illusions perdues*, 1837–43) with Percy a Lucien de Rubempré, who ends on a happier note. *Ranthorpe* also includes a proto-Chernyshevskyan medical student, a "mixture of the gentleman and the Mohock" whose "dark eye was full of fire and intelligence; his open laughing face was indicative of malicious mirth and frankness; and the resolution about his brow, and sensibility about his mouth, redeemed his slang appearance, and showed the superior being, beneath the unprepossessing exterior" (Lewes, *Ranthorpe* 6). Harry Cavendish's first act in the novel is to knock to the ground a peddler who "was beating his donkey in so brutal a manner that several people were crying 'Shame! shame!'"; in a chapter prefaced by an epigraph from George Sand's *Jacques* (1834), Harry ends by breaking his engagement with Isola when he realizes that she loves Percy instead (7). In between, Harry solves a violent murder wrongly attributed to Percy, and it is in this "sensational" subplot that Lewes's pioneering science of mind and body comes into play.

Not unlike Porfiry Petrovich in *Crime and Punishment*, Harry's strategy in the absence of any hard evidence is simply to confront the real murderer, Oliver Thornton, who, despite a plan that "seems to succeed even in its smallest details" (Lewes, *Ranthorpe* 208), is overwhelmed with guilt. In the narrator's words, "[h]e had thought of flying to America, but was afraid, lest it should look suspicious.... Such was his suffering, that he was often on the point of blowing his brains out, and so ending his misery" (222); "[e]very knock at the door went to his heart, as if it announced his arrest. Every noise in the street sounded like the mob coming to seize him. He read the morning and evening paper with horrible eagerness. Every line respecting the murder made him thrill" (221). The "thrill" of Oliver's response is as physical as it is mental, as is the initial motivation for his crime. The narrator explains:

> His uncle's death soon became a fixed idea with him.... He must either become a murderer or a monomaniac! The tyrannous influence of fixed ideas – of thoughts which haunt the soul, and goad the unhappy wretch to his perdition – is capable, I think, of a physiological no less than of a psychological explanation ... In proportion to the horror or interest inspired by that thought, will be the strength of the tendency to recurrence. The brain may be then said to be in a state of partial inflammation, owing to the great affluence of blood in one direction. And precisely as the abnormal affluence of blood towards any part of the body will produce chronic inflammation, if it be not diverted, so will the current of thought in excess in any one direction produce monomania. Fixed ideas may thus be physiologically regarded as chronic inflammations of the brain. (202–3)

Although a recent strain of scholarship, as Rylance notes, tries to enlist Lewes for the Kantian side, Lewes's stance in *Ranthorpe*, as in his later scientific writings, is not the idealism that "the tyrannous influence of fixed ideas" taken by itself might suggest (Rylance 327–30). Nor, despite the emphasis on the material body with its "chronic inflammations" and "affluence of blood," was Lewes an advocate of the more radical nihilist approach that would elevate body over mind. In the novel, as in the later scientific work, body and mind instead operate in tandem, as simultaneously both cause and effect.

In his 1876 article "What is Sensation?," for example, Lewes addresses the ambiguity of a term that was used by physiologists and psychologists alike in different meanings. The physiologist, he explains, takes the word in its "objective aspect" to refer to "the reaction of a sensory organ" alone, as separate from our conscious recognition of that reaction (Lewes, "Sensation" 159). The psychologist, on the other hand, "has only direct knowledge of a change of feeling following some other change," a very different kind of "sensation" that is "wholly a fact of Consciousness" (159). While Lewes suggests that scientists might do better to use "sensation" for what he calls "neural processes" and "feeling" for "psychical" ones, his clarification comes only with the caveat that the two are ultimately interdependent. As Lewes writes in the first volume of his *Problems of Life and Mind* published in the same year, "every mental phenomenon has its corresponding neural phenomenon (the two being as the convex and concave surfaces of the same sphere, distinguishable yet identical), and ... *every neural phenomenon involves the whole Organism*" (Lewes, *Problems* 1: 103–4). The "thrill" that both Oliver Thornton and his reader experience offers an early illustration of the theory of mind and body as simultaneously both one and two that Lewes would develop over his career in science. That science would find a still more significant literary expression in a genre that in 1847 had yet to really appear on the scene: the novel of sensation.

As Richard Fantina argues, "sensation" is most often defined in terms of the lurid plots that mark the novels that appeared after the enormous success of *The Woman in White*, including most famously Mary Elizabeth Braddon's *Lady Audley's Secret* (1862), *Aurora Floyd* (1863), and *The Trail of the Serpent* (1864; first published as *Three Times Dead*, 1860) and Mrs. Henry (Ellen) Wood's *East Lynne* (1861). This high level of plot-tiness also earned the genre almost unmitigated critical disdain. While readers across Europe read and enjoyed their work – in Russia, none other than Tolstoy expressed his admiration for Braddon and especially Wood – the three were also widely accused of making recourse to a "low" kind of plot in order to achieve a commercial "sensation" of the sort that other, more "serious" writers could only envy. As George Eliot wrote in an 1866 letter to her publisher John Blackwood:

> I sicken with despondency under the sense that the most carefully written books lie, both outside and inside people's minds, deep undermost in a heap of trash. I suppose the reason my 6 / [shilling] editions are never on the railway stall is partly [… because] they are not so attractive to the majority as *The Trail of the Serpent*. (Eliot, *Letters* 4: 309–10)

If "sensation" indicates enormous popular success as well as a particularly involved sort of plot, in its own day the term was invented above all to describe the physiological response that its creators apparently intended to elicit.

Where Beer associates the novel of sensation with a Darwinian "appeal to the physical and the material," the overwhelming critical tendency has been to lump plot, popularity, and physiological effect into one "vulgar" whole. As Nicholas Daly writes, critics of the day understood the novel of sensation as not just emerging from the claims of "scientific" materialism but also as attempting to capitalize on its insights in order to "conjure up a corporeal rather than a cerebral response in the reader" (Daly 40). This physical effect was held to be achieved in readers' bodies not just in response to a particularly dramatic plot but also by way of a sympathetic identification with the characters' own often fraught physical states. In her 1862 review of Collins's novel, for example, novelist Margaret Oliphant (1828–97) marvels at the effect produced when the "Woman in White" reaches out to touch Walter's shoulder: "Few readers will be able to resist the mysterious thrill of this sudden touch. The sensation is distinct and indisputable. The silent woman lays her hand upon our shoulder as well as upon that of Mr Walter Hartright." Noting that the effect is then repeated when Walter makes the connection between his chance companion and Laura, Mrs. Oliphant concludes: "These two startling points of this story do not take their power from character, or from passion, or any intellectual or emotional influence. The effect is pure sensation, neither more nor less" (Page 119). Only a year later, the Rev. Henry Mansel (1820–71) argues, sensation had become the marker of all current British writing:

> A great philosopher has enumerated in a list of sensations "the feelings from heat, electricity, galvanism, &c.," together with "titillation, sneezing, horripilation, shuddering, the feeling of setting the teeth on edge, &c."; and our novels might be classified in like manner, according to the kind of

sensation they are calculated to produce. There are novels of the warming-pan, and others of the galvanic-battery type – some which gently stimulate a particular feeling, and others which carry the whole nervous system by steam. There are some which tickle the vanity of the reader, and some which aspire to set his hair on end or his teeth on edge; while others, with or without the intention of the writer, are strongly provocative of that sensation in the palate and throat which is a premonitory symptom of nausea. (Mansel 487)

Like Mansel, most reviewers were highly uncomfortable with the notion that what the sensation novel produced was a kind of "thinking without thinking," what *The Christian Remembrancer* in 1864 described as a "drop from the empire of reason and self-control … which is a consistent appeal to the animal part of our nature" (Maunder 108). As the last chapter has already argued, however, Mansel et al. seem to have mistaken Collins's intent.

Certainly, bodies matter to Collins, as they do to Lewes. In *The Woman in White*, it is not just Walter whose "every drop of blood … was brought to stop by the touch of a hand," but Marian whose body turns "hot and cold alternately" (Collins, *Woman* 15, 262). In *The Moonstone*, both the mystery and its eventual solution revolve around an organism disarranged by a dose of opium combined with a bout of tobacco withdrawal–induced insomnia. As Mrs. Oliphant's review rightly suggests, Collins also deliberately constructed his novels around shocking revelations, including "The Narrative of the Tombstone" in the former and Rachel's stunning accusation in the latter: "*'You villain, I saw you take the Diamond with my own eyes!'*" (Collin, *Moonstone* 352). The comparison with Lewes, however, makes especially clear the distinction between Fosco and his creator that I have already argued. Where Fosco supports the critics' claim that minds are ruled by bodies, Collins himself makes the Lewesian case that both mind and body are implicated in a material world that comes into being only through the medium of our own perceptions.

As his endorsement by the likes of Lebeziatnikov would suggest, Lewes was well known for his focus on the functioning of spinal cord and somatic nerves. As he writes in *The Physiology of Common Life*, and as Franklin Blake's adventures would illustrate, "the Brain is only one organ of the Mind, and not by any means the exclusive centre of Consciousness" (Lewes, *Physiology* 2: 11). Lewes was also, as Rylance writes, "unusually aware of psychological dysfunction and abnormality," and he had a clear sense of the possible physiological bases of any mental disturbance (Rylance 305). In *Problems of Life and Mind*, he suggests that we "take an example from Insanity":

> A visceral disturbance, especially in the digestive or the generative organs, will cause a perversion of Sensibility from which will arise abnormal sensation, hallucination, moods, melancholy, depression, etc. These prompt the intellect to explanation. External causes are imagined; and the wildest hypotheses of persecution, divine or diabolic communication, are invented. As the disturbance spreads and the organism becomes more and more abnormal, the ideas become more and more incoherent, till Dementia supervenes. (Lewes, *Problems* 1: 113)

Even so, as Rylance argues, for Lewes "it is reductive and misleading to claim that all our knowledge is generated from sense experience (though it is ultimately referable to it)" (Rylance 296). Exactly the same caveat holds true for Collins, this despite the emphasis that his novels place on the workings of our senses.

Leaving aside what would seem the obvious fact that novel-reading can impact readers' bodies only by way of the cognitive work that reading requires, it is because sensory experience is always incomplete that Rachel could see what she thought she saw, or that Marian under the influence of Fosco's magnetic personality could dream of Walter. In Lewesian terms, a confusion of "Objective-Subjective and Subjective-Objective" is inevitable, as the dual workings of mind and matter not only complicate our understanding of cause and effect but also guarantee the impossibility of a single objective reality "out there." What Lewes calls his "Reasoned Realism" in fact collapses any distinction between objects as they are and objects as they seem to be. In Lewes's argument, the "senses don't directly apprehend – or *mirror* external things." Instead, "[e]ach excitation has to be *assimilated*," first in terms of the material reality of our particular perceptual apparatus, and then as a reflection of the subject's own evolving history (Lewes, *Problems* 1: 113). "What the Senses inscribe on [the mind]," Lewes writes, "are not merely the changes of the external world; but these characters are commingled with the characters of preceding inscriptions. The sensitive subject is no *tabula rasa*; it is not a blank sheet of paper, but a palimpsest" (1: 149). Collins's many narrators – eleven in *The Woman in White* and eleven again in *The Moonstone*, each telling only as much of the story as he or she knew at that time – offer Collins's broad expression of the idea that Lewes puts much more succinctly: "objective existence *is* to each what it is felt to be" (1: 175).

Although reality in Lewes as in Collins is the product of minds and bodies working together, that reality still enjoys a material existence of its own. Again, "the external world exists," Lewes writes, "and *among* the modes of its existence is the one we perceive" (Lewes, *Problems* 1: 168): perception as mind as well as body is only one part of a material world that is always multiply expressed. Certainly, our perceptions can be wrong, as in Lewes's "example from Insanity," Rachel's accusation, or even, as it turns out, in our insistence on the empirical validity of Euclidean assumptions. In the contemporary terms of the Santiago school of cognition, sensation as always both nerves and neuroses nonetheless functions neither to represent an "objective" reality nor to invent a reality of its own. As Francisco Varela writes, it is rather that "animal and environment are two sides of the same coin, knower and known are mutually specified" (Maturana 253). To make the same point in Latourian terms, for Lewes as for Collins, phenomena "are not found at the *meeting point* between things and the forms of the human mind." They are instead "what *circulates* all along the reversible chain of transformations" that makes us one with the material world (Latour 71).

With his affinity for a twentieth- and twenty-first-century science that sets minds in bodies and the two together in the world as two aspects of a single whole, Lewes, like William James, might seem an exception in a nineteenth century otherwise committed to what Levine calls "the almost masochistic objectivist model," in fact, the same exception: as Rylance points out, James is well known to have read Lewes (Rylance 105). If they are less familiar to readers today, however, in their own day neither Lewes nor the novel of sensation were the outliers that their critics would have us believe. The enormous popularity that Lewes like the novelists of sensation enjoyed would seemingly all on its own argue the importance to the nineteenth century of a more complicated kind of materiality, although we can only qualify any such claim; unfortunately, as the examples of both the real Rev. Mansel and the fictional Lebeziatnikov would suggest, the real nuance of their science was evidently lost on many of their readers. More telling is Beer's contention that Dickens and even Gaskell drew on a Darwinian "emphasis on substantiation" in order to achieve a physical effect in their readers. To associate these physical effects not just with Darwin but also with Lewes is then to extend her argument to include still more nineteenth-century literature of an equally canonical sort. Once we cast sensation in Lewesian terms, we account for its presence in the work of Lewes's closest intellectual collaborator, Marian Evans, a.k.a. George Eliot. Lewes also lays bare the workings of sensation in another writer more often celebrated for his cerebral effects alone: Dostoevsky.

Eliot and Dostoevsky, I: Plot and Plot-tiness

Eliot herself apparently precluded any consideration of her work in sensational terms, including in her sharp rejection of a certain kind of plot and plot-tiness in the famous essay that she published anonymously in the *Westminster Review* in 1856, "Silly Novels by Lady Novelists." As Gordon Haight notes, Eliot started work on her first foray in fiction-writing, *Scenes from Clerical Life* (1857), a mere two weeks after completing "Silly Novels." That Marian Evans would so shortly emerge as the novelist George Eliot only makes especially clear the personal dimensions of the criticism that her essay offers (Haight 210). As the title promises, Eliot categorizes the work of "Lady Novelists" by the "particular quality of the silliness that predominates in them – the frothy, the prosy, the pious, or the pedantic" (Eliot, *Essays* 301), with particular attention to a large group that manages to combine all four into what she calls "a composite order of feminine fatuity" (301). In the case of the recently published *Compensation: A Story of Real Life Thirty Years Ago* (1856), what Eliot also calls the "mind-and-millinery species" entails "a wonderful *pot pourri* of Almack's, Scotch second-sight, Mr. Rogers's breakfasts, Italian brigands, death-bed conversions, superior authoresses, Italian mistresses, and attempts at poisoning old ladies, the whole served up with a garnish of talk about 'faith and development,' and 'most original minds'" (307). By 1866 and the letter to her publisher quoted above, Eliot was already a "superior authoress" herself, and the kind of "Lady Novelist" that she wasn't was Mary Elizabeth Braddon, as "silly" had given way to "sensational."

Of course, as early as 1859 and *The Woman in White*, any "talk about 'faith and development'" had already gone by the wayside. Collins was also not a lady, although many of the novelists of sensation were. Collins's complicated and often lurid intrigue, replete with second-sight, an Italian count, and real murder, would nonetheless seem exactly the kind of plot that Eliot pledged herself not to write. While a great many readers have taken Eliot at her word, a more recent strain of scholarship questions her claim to another sort of writing. In "Strange Sympathies: George Eliot and the Literary Science of Sensation" (2013), Mary Beth Tegan's focus is the "interest in the blurred edges of sympathy and sensation" (Tegan 172) that Tegan identifies in Eliot's 1862 novel *Romola* and that she associates with Eliot's reading not just of Lewes, but of Herbert Spencer and of Alexander Bain's "Literature of Plot-interest" (1859). Tegan also draws our attention to the "highly physical language" that marks Eliot's letter to her publisher Blackwood, not just in terms of the despondency that "sickens," but especially in Eliot's immediately subsequent attempt to downplay the significance of the physical response that she has just described (Tegan 170). "My mind is not much molested by these things," Eliot continues, "but writing to you, I am betrayed into this gossip about my authorship's affairs" (Eliot, *Letters* 310). As Tegan acknowledges, she is not the first to remark on Eliot's own attempts to "molest" readers' minds. "Perhaps the most trenchant comment about Eliot's sensationalistic tendencies," Tegan writes, "was made by Charles Reade, who claimed that *Romola* 'borrowed' from his own fictional return to the fifteenth century, *The Cloister and the Hearth* (1861)" (Tegan 171). No doubt stung by Eliot's criticism of his own novelistic "excess" in her review of *It Is Never Too Late to Mend* (1856), published in the same year as "Silly Novels," Reade responds in kind: "George Eliot is a writer of the second class," he writes, "adroit enough to disavow the sensational, yet to use it as far as her feeble powers would let her" (Tegan 171). Tegan also notes the now common reading of Eliot's novella *The Lifted Veil* as a species of sensation.

Where *Romola* is among the less well known of Eliot's works, *The Lifted Veil* is most often regarded as an active embarrassment. *The Lifted Veil* is the first-person account of an experience of sustained clairvoyance. Cursed with a "lifted veil" that allows him insight into the future as well as the thoughts of the people around him, Latimer falls in love with the one person whose mind is closed to him, his brother's fiancée, Bertha. While Latimer succeeds in marrying Bertha after the sudden death of his brother, the marriage fails as he comes to recognize Bertha's real nature and as his own health declines. At the end of the story Latimer is visited by a friend from his youth, a certain Charles Meunier, now a scientist of European renown. When Bertha's maid promptly dies under suspicious circumstances, Meunier proposes an experiment, an immediately posthumous blood transfusion that restores Mrs. Archer to life only long enough to accuse Bertha of the slow poisoning of Latimer. For most readers, the plot reeks more of Poe than it does Eliot. Despite the contemporary (pseudo-)science that Eliot brings to bear, the combination of poisoning, deathbed conversion, and what I can only call "Scotch second-sight" also offers a striking example of exactly the sort of "silliness" that Eliot affects to deplore. Kate Flint restores this admittedly difficult work to our consideration, however, by reading it side-by-side with Lewes's *Physiology of Common Life*. As Flint writes in "Blood, Bodies and *The Lifted Veil*" (1997), *The Lifted Veil* on the level of form as well as content dramatizes Lewes's concerns with "the relationship of the normal and the abnormal, the connections between mind and brain and between cerebral activity and physical functions" (Flint, "Blood" 471). The same is also true for *Middlemarch*.

As Flint argues, *The Lifted Veil* problematizes a Victorian fascination with visibility that Menke in "Fiction as Vivisection: G. H. Lewes and George Eliot" (2000) extends to include not just Lewes, but also Eliot's later and much more canonical work, *Middlemarch*. In Menke's reading, Latimer's clairvoyance and Meunier's experiment stand in for the aims of Eliot's fiction writing as well as for the theory and practice of vivisection advocated by both the real Lewes and the fictional Lydgate.[6] Latimer's plight, he writes, is summed up by the narrator's famous lines in *Middlemarch*: "If we had a keen vision and feeling of all ordinary human life, it would be like hearing the grass grow and the squirrel's heart beat, and we should die of that roar which lies on the other side of silence" (Eliot, *Middlemarch* 194).[7] I would only add that *Middlemarch* is finally like *The Lifted Veil* in its surprising, if subdued, reliance on the tropes of sensation, including, to borrow again from Fantina, the themes of inheritance, poisoning, drug abuse, and adultery.

In the case of Will and Dorothea as well as Will and Rosamund, the adultery doesn't quite happen, although their fellow Middlemarchers are sometimes astonishingly quick to worry that it will. In the case of old Peter Featherstone, it evidently did, as the arrival of his hitherto unacknowledged illegitimate son, the "frog-faced" Joshua Riggs, upends his family's expectations of inheriting the Featherstone estate. Murder also hovers just off-stage, most significantly in the case of Raffles, whose death implicates both Bulstrode and also Lydgate in a case of deliberate malpractice that involves its own sort of poisoning. Murder in *Middlemarch* is also specifically a female response to unwanted men. Lydgate is speaking only metaphorically when he describes Rosamund as his "basil plant"; basil, he explains, "flourishe[s] wonderfully on a murdered man's brains" (Eliot, *Middlemarch* 835). His first love, Laure, however, really did stab her husband to death in an apparent accident on stage that she later acknowledges was no accident at all. We might finally note the implication of Dorothea in the death of her first husband, Casaubon. As Dorothea's many fans will rush to point out, her much older husband came to the marriage with a weak heart. Still, Casaubon's death frees Dorothea to eventually marry Will, a prospect that Dorothea's family and friends, if not her readers, view pointedly because anachronistically in "sensational" terms. As Mrs. Cadwallader says and Dorothea repeats two pages later, marrying Will would be like marrying "an Italian with white mice" (Eliot, *Middlemarch* 490). While Mrs. Cadwallader's point of comparison in her fictional 1830s is a little unclear, by 1871 and the publication of *Middlemarch*, there is only one Italian with white mice: Count Fosco. The steady undertow of sensation that runs all through Eliot's plot is also made much more overt in her characterization.

As again Flint argues in "The Materiality of *Middlemarch*" (2006), just as Eliot's letter to her publisher John Blackwood with its "sickening" despondency and mental "molestation" lends thought a striking degree of materiality, so Raffles arrives in the novel as the embodiment of Bulstrode's "incorporate past," a "loud red figure ... risen before him in unmanageable solidity" (Eliot, *Middlemarch* 523). In *Middlemarch* Eliot also dwells on the physical effects of ideas and emotions. In the case of Bulstrode it is the impact of the mind on the material world that seems most evident, not just in his apparent manifestation of his shameful past in the actual person of Raffles but also in the effect of the subsequent mental strain on his physical being. Without understanding the reasons for Bulstrode's decline, Lydgate reflects that "One sees how any mental strain, however slight, may affect a delicate frame" (680). When Bulstrode's past misdeeds are finally known to all, his wife, too, is marked by "the sorrow that was every day streaking her hair with whiteness" as by a face "aged to keep sad company with his own withered features" (824, 825). Eliot consistently lays emphasis on the workings of the body, however, when the narrator describes Bulstrode's "diseased motive" as operating "like an irritating agent in his blood" (707–8), or when she compares his "misdeeds" to "the subtle muscular movements which are not taken account of in the consciousness" (687). The centrality of the body is still more evident in the case of Will.

Will, the narrator tells us, "was made of very impressible stuff," and his body and mind are one, for example, in his habit of "shaking his head backward" in conversation, or in an emotional moment with Dorothea when "the blood had mounted to his face and neck," or especially in the reaction of his nerves that Eliot repeatedly describes as "electric shock" (Eliot, *Middlemarch* 388, 364, 631, 543). When Dorothea enters the room, for example, Will "started up as from an electric shock and felt a tingling at his finger-ends." "Any one observing him," the narrator adds, "would have seen a change in his complexion, in the adjustment of his facial muscles, in the vividness of his glance, which might have made them imagine that every molecule in his body had passed the message of a magic touch" (388). When much later in the novel Rosamond lays the tips of her fingers on Will's coat sleeve, an infuriated Will recoils from her touch, "darting from her, and changing from pink to white and back again, as if his whole frame were tingling with the pain of the sting" (777). As Flint writes, Eliot's intent is not to show bodies operating without minds. Eliot makes the point instead "not only that body and mind are inseparable but that the properties of the mind itself depend on the materiality of the body" (Flint, "Materiality" 79). The same is also true in *Crime and Punishment*.

If the comparison of Dostoevsky with Eliot strikes readers as a little surprising, it is largely because Dostoevsky's reliance on "sensational" twists and turns of plot is anything but subdued. In contrast to Eliot's restraint in *Middlemarch*, not just *Crime and Punishment*, but all the novels post-Siberia, from *The Insulted and Injured* through *The Brothers Karamazov*, tell stories of not just possible, but real (if complicated) murder, prostitution, adultery, and illegitimacy. His investment in a "stubborn" materiality, however, is exactly hers. In "Sensual Mind: The Pain and Pleasure of Thinking" (2004), Marina Kostalevsky vividly evokes the materialization of mental processes as characteristic of Dostoevsky's writing as it is of Eliot's. In *Notes from Underground*, the unnamed narrator gives thought a physical existence when he writes: "A sullen thought was born in my brain and passed through my whole body like some vile sensation, similar to what one feels on entering an underground cellar, damp and musty" (*Notes* 88; *PSS* 5: 152). In *The Brothers Karamazov*, chapter titles like "The Sensualists" ("Сладострастники") and "An Adulterer of Thought" ("Прелюбодей мысли") suggest not just a physicality, but even an "eroticism of intellectual desires" (Kostalevsky 202). As the far greater prominence accorded to drunkenness, dirt, and attacks of brain fever makes only more apparent, Dostoevsky also pays pointed attention to the workings of his characters' bodies.

In contrast to Eliot's openly omniscient narrator, the narrator of *Crime and Punishment* confines himself almost entirely to Raskolnikov's point of view, so much so, in fact, that when Raskolnikov faints at the police station, the narrative stops: "Raskolnikov picked up his hat and made for the door, but he did not reach it … When he came to his senses he saw that he was sitting in a chair" (*Crime* 105, ellipses in the original; *PSS* 6: 83). Despite the excess of Dostoevsky's plot, we are perhaps most immediately struck by the psychological dimensions of that perspective: before the abrupt shift in style that begins with the confession in the final lines of Part Six, what we see pours out in a dense and hectic flow of words that corresponds to Raskolnikov's dense and hectic mind. As Sarah J. Young argues, however, "the interiority for which … [Dostoevsky] … is so famous does not deny physical being, but to the contrary implies the existence of an exterior" (S. Young 119). In *Crime and Punishment*, that exteriority includes the city of St. Petersburg as if plotted on a map as well as the many rooms that make up Raskolnikov's seedy neighbourhood. As Dostoevsky's many descriptions of his physical state would remind us, Raskolnikov's narrating mind is also bounded by a body.

The text in fact devotes significant space to accounts of Raskolnikov's sleeping and especially to his often "greedy" consumption of soup, bread, tea, and beer; it also dwells on the corporeal effects of an illness that is as physical as it is mental. As Zosimov sees it, Raskolnikov's illness is the "product of many complex moral and material influences" (*Crime* 207; *PSS* 6: 159), while Raskolnikov himself distinguishes body from mind while nonetheless suggesting their interaction when he reflects on conditions at the police station: "'too bad it's so airless here,' he added, 'stifling … My head is spinning even more … my mind, too'" (95; *PSS* 6: 75). Porfiry Petrovich also emphasizes Raskolnikov's corporeal self, for example, when he recalls the physical effects of an earlier meeting: "your nerves were humming and your knees trembling, and my nerves were humming and my knees trembling" (449; *PSS* 6: 313), or when he explains Raskolnikov's return to the scene of the crime as a desire to again experience "that spinal chill" (456; *PSS* 6: 346). The body is also at issue in Raskolnikov's repeated experience of "sensations" [ощущения].

When Raskolnikov realizes that the police have called him in to question him not about the murder but about the money he owes, he is at first filled with "complete, spontaneous, purely animal joy" (*Crime* 98; *PSS* 6: 78). This unthinking emotion quickly gives way, however, to something much more troubling:

A dark sensation of tormenting, infinite solitude and estrangement suddenly rose to consciousness in his soul.… What was taking place in him was totally unfamiliar, new, sudden, never before experienced. Not that he understood it, but he sensed clearly, with all the power of sensation, that it was no longer possible for him to address these people in the police station, not only with heartfelt effusions, as he had just done, but in any way at all.… Never until that minute had he experienced such a strange and terrible sensation. And most tormenting of all was that it was more a sensation than an awareness, an idea; a spontaneous sensation, the most tormenting of any he had yet experienced in his life. (103–4; *PSS* 6: 81–2)

Once he hides his ill-gotten gains, this "new, insurmountable sensation" becomes "a certain boundless, almost physical loathing for everything he met or saw around him" (110; *PSS* 6: 87), a physical response that is only entirely sorted out at the very end of the novel when a different "sensation" [ощущение] "seized him all at once, took hold of him entirely – body and mind," and he bows down at the crossroads to kiss the earth (525; *PSS* 6: 405).

For Kostalevsky, Dostoevsky's "sensualization of the mind" derives from "the concept of unity between mind and body as developed by the Orthodox tradition, with its attention to the sacredness of matter" (Kostalevsky 207). As P.S. Shkurinov argues, however, for all the centrality of Dostoevsky's faith in an intangible God, Raskolnikov's finally uncontrollable urge to confess is also a physical reaction, the response of "living life" to the deed Raskolnikov's rational mind has led him to perform (Shkurinov 230). Indeed, as the example of the very English and eventually nearly atheist George Eliot would also suggest, terms like "spinal cord" [спинной мозг] and "sensation" [ощущение] point straight to a scientific context fascinated by the possibilities of a central nervous system and bodily responses that bypass cognition. While Dostoevsky's reputation as a writer of intellectual heft and difficulty, like Eliot's, has long obscured his "sensational" underpinnings, his first readers were quick to criticize his reliance on corporeal effects. In Great Britain the attack on sensation came from the right, in Henry James's dismissal of *Middlemarch* as "too often an echo of Messrs. Darwin and Huxley" (H. James 428) as in Rev. Mansel's "horripilation." In the Russian context, the same criticism was directed from the left, as in Pyotr Tkachev's 1873 review of *Demons*, "Sick People" ("Больные люди").[8]

While his earliest writings appeared in Dostoevsky's journals *Time* and *Epoch*, Tkachev (1844–86) quickly joined the nihilist camp in literary as well as political terms. By the late 1860s, he was involved with Sergei Nechaev, the radical left-wing activist whose organization of a terrorist cell around a communal act of murder served as Dostoevsky's starting point for the fictional *Demons*. After a brief stint in prison, Tkachev emigrated in 1871 and resumed his journalistic career abroad. Like his mentor Chernyshevsky, Tkachev combined a would-be material determinism with a paradoxical insistence on maintaining and even extending intellectual control, and in his review of *Demons*, he faults Dostoevsky on the latter score. For Tkachev, the "sick people" of his title are first Dostoevsky and then his characters, all of whom he sees as suffering from a sort of schizophrenia. His real concern, however, is for the reader, who apparently suffers in Russia as in Great Britain from the new literature of sensation. Dostoevsky's writing, Tkachev argues, reflects an impoverished literary environment so desperate for "nervous irritation [нервного раздражения]: scandals, horrors, piquancy" that it makes recourse to "police agents, examining magistrates, and even just district court stenographers" (Tkachev 75–6). Tkachev is appalled at what he describes as Dostoevsky's method in *Demons*: "Give us more and more gossip, scandal, irritate all the more strongly the reader's spinal cord [спинной мозг], make his hair stand on end, entertain him, amuse or frighten him, but just don't make him think or look up from the page" (Tkachev 75). In his influential article "A Cruel Talent" ("Жестокий талант," 1882), the only slightly less radical Nikolai Mikhailovsky (1842–1904) makes the same point with regard to Dostoevsky's oeuvre as a whole.

Unlike Tkachev, Mikhailovsky was never forced to flee Russia, nor did he serve time in Russian prison. If his views were marginally less extreme, however, he was nonetheless cut from the same nihilist cloth. Mikhailovsky is well known for his contributions to the "Woman Question," including an enthusiastic preface to the Russian translation of John Stuart Mill's *Rights of Women* in 1869, and he was an active member of the revolutionary political organization "The People's Will." He also shared Tkachev's misgivings regarding Dostoevsky. While he grants Dostoevsky's formal ability, Mikhailovsky also sees a deliberate and sustained attempt to inflict suffering on the reader through the use of "excessive and entirely inartistic longeurs, introductory scenes ... [and] ... digressions" (Mikhailovskii 332). This "cruel talent," he continues:

> will cloud your mind with its images and pictures and make your heart beat faster, and only in those lucida intervalae when in the course of reading sobriety returns to us, will you ask yourself: Why is he so tormenting that Sidorov or Petrov? Why is he is titillating [щекочет] me, too, in such tormenting fashion?

In fact, Mikhailovsky explains, there is no purpose to this suffering other than to create "sensations [ощущений] that become a need" (333), as in his estimation Dostoevsky's writing served Russian society of his day as nothing more than a kind of "narcotic" [наркотического свойства] (334). The echo is so exact that it seems hardly likely that Mikhailovsky didn't already know the English criticism of sensation. In the words of the *Christian Remembrancer*, sensationalism "devotes itself to harrowing the mind, making the flesh creep, causing the hair to stand on end," a goal that it achieves "by drugging thought and reason" (Maunder 107). The misunderstanding is also the same: like Collins, Eliot, and Lewes, Dostoevsky advocates not mind over body, but their mutual interaction.

Dostoevsky famously shares Lewes's fascination with "psychological dysfunction and abnormality," and, like Lewes, he persistently raises the possibility of a physiological basis. In *Crime and Punishment*, psychological states altered to the point of insanity abound, almost always with an often ambivalent reference to a physical state. Just as Raskolnikov's own illness partakes of mind and body alike, so Katerina Ivanovna's eventually full-blown insanity is linked to her consumption. Marmeladov implies that her poor grip on reality precedes her actual illness when he attributes a kind of innocence to the lies that his wife would tell on his behalf: "she believed it all, she delights in her own fancies, by God, sir!" (*Crime* 21; *PSS* 6: 19). Katerina Ivanovna's nervous energy may also have produced the consumption, as the description given when Raskolnikov first sees her would suggest: "Her eyes glittered as with fever, but her gaze was sharp and fixed, and … this consumptive and agitated face produced a painful impression" (25; *PSS* 6: 22). Physical and mental are finally inextricably connected in Lebeziatnikov's attempt to explain Katerina Ivanovna's deranged behaviour in the wake of her husband's funeral.

While Lebeziatnikov himself acknowledges that he knows nothing about medicine, he is quick to make a diagnosis. He first tells Raskolnikov, "She's certainly gone mad!… It's those little knobs they say come out on the brain in consumption," only to apparently dismiss his first theory for another: "But what I say is this: if one convinces a person logically that he essentially has nothing to cry about, he'll stop crying." If a nihilist might seem an unlikely proponent of a talking cure, still more striking is the commitment to the physiological underpinnings of mental states evident in Raskolnikov's response. "Life would be too easy that way," Raskolnikov replies, prompting Lebeziatnikov to give a somewhat confused account of a Parisian professor's success in curing mad people by "logical conviction" alone. Lebeziatnikov explains:

> His basic idea is that there's no specific disorder in a mad person's organism, but that madness is, so to speak, a logical error, an error of judgment, a mistaken view of things. He would gradually prove his patient wrong, and imagine, they say he achieved results! But since he used showers at the same time, the results of the treatment are, of course, subject to doubt … Or so it seems" (*Crime* 423–4; *PSS* 6: 325).

Both the ambivalence and the return always to the possible physiological bases of mental derangement mark also the many "perversions of sensibility" in *Crime and Punishment* that stop just short of insanity.

Svidrigailov, for example, claims to see ghosts, and significant again is Raskolnikov's response: "You should see a doctor" (*Crime* 288; *PSS* 6: 220). Wrong or at least "unreal" perceptions are also offered by the dreams that again apparently stem from particular physical conditions. Raskolnikov's dream of the mare, for example, derives from the "morbid condition" that, as the narrator explains, tends to give dreams a "remarkably graphic, vivid, and extremely lifelike quality" (54; *PSS* 6: 45).

In the same way, according to Nastasya, Raskolnikov's still more life-like dream that the police lieutenant was beating his landlady can be explained by "the blood." No one was here, she says: "It's the blood clamoring in you. When it can't get out and starts clotting up into these little clots, that's when you start imagining things" (117; *PSS* 6: 92). We might consider finally how often Raskolnikov is taken to be drunk. As Tatiana Kasatkina argues, this misinterpretation offers a "real" reading of what is actually a spiritual condition, that he has "regaled … [his] … spirit … with the drunkenness of sin" (Kasatkina 13). This persistent mistake also, however, reflects a clear understanding that chemical imbalances in the brain can also be artificially induced. In *Middlemarch* Mr. Farebrother makes the same mistake when he sees a visibly agitated Lydgate: "'He may have been taking an opiate,' was a thought that crossed Mr Farebrother's mind" (Eliot, *Middlemarch* 640).

There is always the possibility in Dostoevsky that apparently "abnormal" perceptions are not mad at all, but instead windows onto alternate realities. Despite Raskolnikov's questioning of her sanity, for example, Sonia is not mad but rather sees a Christian meaning to her life that Raskolnikov eventually comes to share, and we might also note Svidrigailov's answer to Raskolnikov: "I agree that ghosts come only to sick people; but that only proves that ghosts cannot appear to anyone but sick people, not that they themselves do not exist" (*Crime* 289; *PSS* 6: 221). While readers who often don't share Dostoevsky's religious beliefs are quick to associate alternate realities with supernatural beings, multiple realities are also a direct consequence of the grounding of perception in the material workings of the body. In Young's reading of Dostoevsky, dreams like delirium "present a concentration of sensual perception" (S. Young 122); as William James puts it with evident knowledge of Lewes's work and a clear affinity for Dostoevsky's: "for aught we know to the contrary, 103° or 104° Fahrenheit might be a much more favorable temperature for truths to germinate and sprout in, then the more ordinary blood-heat of 97 or 98 degrees" (W. James 21). If Lewes was also drawn to "psychological dysfunction," the larger scientific point is that bodies even in a "normal" state bring a certain perceptual apparatus to bear. As again James writes, "[t]he knower is an actor, and co-efficient of the truth on one side, whilst on the other he registers the truth which he helps to create" (Davis 137). Similarly, Michael Davis argues, for Lewes "neither the mind nor the external world on their own can be the sole arbiters of knowledge: instead, we can only know the world through a combination of the world and the mind" (Davis 137). If I once more emphasize Dostoevsky's scientific underpinnings, however, it is again not to diminish the importance of his faith. It is instead to argue that we read both ways at once.

While allusions to specifically Christian morality are not entirely lacking in *Middlemarch*, Eliot is most often read in terms of a science that itself functions as a kind of morality. As Sally Shuttleworth writes, where Bulstrode's religion proves false, his body proves true, as "[t]he physiological coherence of … [Bulstrode's] … body appears to offer a guarantee of cumulative social order" (Shuttleworth 155).[9] Readers of Dostoevsky, on the other hand, tend to limit their reading to a Christian faith operating as if all on its own. As Dostoevsky himself presented it in his pitch to his publisher Katkov, however, Raskolnikov's terrible urge to confess is a matter of material as well as of spiritual necessity. Dostoevsky's summary of his projected novel concludes: "God's truth, the earthly law make themselves felt, and … [the hero] … ends by being compelled to turn himself in" (Dostoevskii *PSS* 28/2: 137). To borrow from Lewes, to read Dostoevsky in terms of sensation is to understand "God's truth" and "earthly law" as twin modes of our existence, "the convex and concave surfaces of the same sphere, distinguishable yet identical." It is also to turn our attention to the body most importantly at stake in his recourse to plot and plot-tiness: our own.

Dostoevsky's Readers

The literature that would dispense with plot has its own devices for reaching real readers in the world, usually constructed around some variant of direct address. In the case of Tolstoy, it is a matter of a sometimes hectoring tone that picks up over the many pages of his novel, continues through two epilogues and finishes, if we can call it that, in the instructions to the reader that make up the later "A Few Words Apropos *War and Peace*" ("Несколько слов по поводу книги *Война и мир*," 1868). In the absence of plot altogether, the authors of "physiologies" fictional as well as real simply refer to "you." As Aleksey Vdovin argues, a number of Dostoevsky's often unnerving discursive choices in *Notes from Underground* are lifted directly from Ivan Sechenov's *Reflexes of the Brain* (*Рефлексы головного мозга*, 1863), including his narrator's relentless reference to "you." An address to "you," as Vdovin notes, was standard practice in the anatomical theatres and public lectures where pioneering Russian physiologist Ivan Sechenov (1829–1905) made his reputation, and it slips easily into his foray into popular science (Vdovin 108). *Reflexes of the Brain* begins with a polite reminder: "You, dear reader, have surely been present at disputes around the existence of the soul and its dependence on the body," only to turn to the description of his apparently objective work in a sterner series of directives (Sechenov 3). "Cut off the frog's head and throw it on the table," Sechenov writes:

> For the first seconds it is as if paralyzed; but in no more than a minute you see that the animal has recovered and sat down on the table in the same pose that it usually takes on dry land…. Leave the frog alone, or, more precisely, don't touch its skin, and it will sit without moving for an extraordinarily long time. Touch the skin, and the frog will move, and again be still. Pinch it a little more sharply, and it might even make a jump, as if trying to avoid the pain (7).[10]

The authors of literary "physiologies" both before and after *Reflexes of the Brain* largely lack Sechenov's scholarly credentials. Even in the absence of any real experience in the field, however, "you" is often wielded to the same "scientific" effect.

The famous literary compendium *Physiology of St. Petersburg* (*Физиология Петербурга*, 1845), for example, is replete with "you," as when V. Lugansky (Vladmir Dahl) in "The Petersburg Caretaker" ("Петербурский дворник") tells us to "[g]o down about six steps, stop, and take in the thick air and clouds of steam" (Kuleshov 42); in "Petersburg Organgrinders" ("Петербургские шарманщики"), Dmitri Grigorovich begins, "[t]ake a look at the man who is slowly crossing the sidewalk; look attentively at his whole figure (51). As Greta Matzner-Gore argues, the effect is to turn the readers of their nonetheless fictional accounts into "virtual field assistants, participants in the process of observation and data collection" (Matzner-Gore, "Empiricism" 365). Ponge's modernist call for what he mistakenly believes is a new kind of literature in his "Text on Electricity" is again aimed at "you," although Ponge starts by categorizing his reader in the third person: "We are looking at someone, or rather, someone is looking at this book in his hands," Ponge writes. "He has opened it. His eyes are now running over these lines, and, in all probability, he is beginning to wish that he could grasp something" (Ponge 157). As he turns to assist his struggling reader, however, Ponge again moves to direct address: "And now, we are going to turn off the ceiling lights on this book and on our intentions and instead, turn on the desk lamps or the bedside ones, and, with your permission, we are going to become more intimate and have a more familiar conversation, in a slightly lower voice" (161). In *What Is to Be Done?* Chernyshevsky's aspirations to the "scientific" rely on the same blurring of apparent objectivity and overt control.

While Chernyshevsky tends to reserve "you, the public" for the readers that he purports to admire, the "you" who "have only just begun to appear among us" (Chernyshevsky, *What* 49), as he feels the need for a more active manhandling of his reader, he shifts to the third person. "Oh, I know," Chernyshevsky's narrator begins, only to interrupt himself, "What's that? A familiar voice…. I turn around to look…. Sure enough! It's he, the perspicacious reader, recently banished in disgrace for his total ignorance of the ABC's of art." As the unfortunate reader tries again, "Oh, I know who wrote that," the narrator puts a stop to his unwanted confidence: "So I seize a napkin," he says, "and stuff it in his mouth" (323). Despite this rough treatment, the "perspicacious reader" is still trying to intervene at the very end of the novel: "'Excuse me, excuse me!' says the reader, not just the perspicacious reader, but every reader, growing more stupefied the more he understands" (444). Even as Chernyshevsky increasingly feeds us our own lines, he reserves the final word for himself. "If you don't want to listen now," the narrator concludes, "then of course I must postpone the sequel until you're in a mood to listen. I hope we won't have to wait too long for that day" (445).

The chastened reader who interjects him- and (less often) herself into the text pre-figures the many real readers who made *What Is to Be Done?* an all-time bestseller in nineteenth-century Russia. Indeed, as measured by the many real-life imitations that it spawned – young Russians who embarked on fictitious marriages, for example, or who set up their own social and economic collectives – Chernyshevsky succeeded in corralling real readers at an astonishing rate. Even granting the novel its enormous impact on the course of Russian history, still Chernyshevsky's repeated attempts to call out and control both "you" and then "him," like the primacy that his characters give to theoretical intervention, once more empties his would-be materialism of any real matter at all. While the revolution that Chernyshevsky sought really came to pass in relatively short order, practice often fell short of theory, not just in the later Soviet Union, but even among Chernyshevsky's most devoted acolytes, a real-life failing that Dostoevsky subjected to merciless mockery, not least in the attempt at a Chernyshevkyan open marriage that goes awry in *Demons*.

In *Demons* the all-too-appropriately named Virginsky, the narrator says, "got everything out of books" (*Demons* 31; *PSS* 10:28), including his decision to give way in favour of his wife's lover. In Dostoevsky's comical rendering, however, books (or at least that particular book) can get even an ardent believer only so far. "It was asserted," the narrator continues, "that when his wife announced his retirement, Virginsky said to her: 'My friend, up to now I have only loved you, but now I respect you'"; as the narrator immediately adds, "but it is hardly possible that such an ancient Roman utterance was actually spoken; on the contrary, they say he wept and sobbed" (32; *PSS* 10: 29). Where Virginsky's reading has evidently set his mind and his body at odds, it is Dostoevsky's hope that his own reader will operate on both levels at once. As Beer writes, "we are to laugh and weep as we read: rictus and wetness." While Dostoevsky, like any writer, lays his traps, he also doesn't insist.

In *Notes from Underground* Dostoevsky's unnamed narrator himself reaches for "you" in an astonishing parody and finally dismantling of the "physiological" form of address as not just Sechenov, nor even, as Greta Matzner-Gore argues, Grigorovich, but above all as Chernyshevsky wields it.[11] It's not just that the narrator, especially in the first part, "Underground," uses his insistent address to "you," also "sirs" and "gentlemen," to create an extraordinary effect of hammering away at his reader, but that, as his rant continues, "you" and "I" change places and even elide. Although readers often take the Underground Man's "you, gentlemen" in its historical meaning to refer exactly to Chernyshevsky and his friends, grammatically it is you, except when it seems to really be me, as in:

> I'll explain to you: the pleasure here lay precisely in the two vivid consciousness of one's own humiliation; in feeling that one had reached the ultimate wall; that, bad as it is, it cannot be otherwise; that there is no way out for you, that you will never change into a different person; that even if you had enough time and faith left to change yourself into something different, you probably would not wish to change; and even if you did with it, you would still not do anything, because in fact there is perhaps nothing to change into. (*Notes* 8; *PSS* 5: 102)

As "you" points to me, the narrator himself loses track of who says and thinks what. "And it is then – this is still you speaking" (24; *PSS* 5: 113), he clarifies, or "(this me speaking now)" (25; *PSS* 5: 113), as, entirely against the narrator's intention, the Underground Man and his reader-opponents collapse into one. Despite what "you" might seem to imply, true dialogue in *Notes from Underground* lies elsewhere, first in the work that the reader does to solve this puzzle of narration, and then in the glimpse that Liza offers of a different way of being, one where others are also selves in their own right. In *Crime and Punishment*, as in all the later novels, Dostoevsky dispenses with "you," however complicated, and develops the possibilities of plot instead.

Plot as content in *Crime and Punishment*, as in *Middlemarch*, winds up with a now successful heterosexual romance pointedly cast as a relationship of equals. Just as the Underground Man is trapped by his desire for "domination and possession," so Raskolnikov's "extraordinariness" relies on his belief that the "old crone" was beneath him; Dorothea's mistake with Casaubon is the flip side of the coin, a woman's desire not for a partner but for a superior being. Certainly, Dostoevsky has other ways of rendering the wrong relationship of self to other, the kind where self tries and fails to construct itself only at the expense of the other. As chapter five will argue, however, most often and across all his works, Dostoevsky demonstrates the murderous effects of hierarchical relationships in terms of men's attempts to construct their own manliness at the expense of women. As a man Raskolnikov only breaks from that cycle when he identifies with woman. "I killed myself," he tells Sonia, "not the old crone" (*Crime* 420; *PSS* 6: 322), and so opens the way to a relationship of equality and shared convictions. Plot as form makes the same point, although not always in the same heteronormative terms.

If all talk might seem less than sexy, the address to "you" is persistently cast as a kind of romance. Even Chernyshevsky is at least flirtatious when he suggests that any apparent hostility is really a sign of hidden attraction. "My, how attached we've become, the perspicacious reader and I," the narrator writes, "He insulted me, I threw him out on his ear twice, and still we go on exchanging our innermost thoughts. A secret attraction of hearts?" (Chernyshevsky, *What* 352). Ponge not only proposes an intimate conversation by the light of bedside lamps, but even seduction in so many words. "I am going to speak about myself," he says, "about the person who was asked to write this text and given the job of seducing you" (Ponge 163). Seduction is more convincingly at work in sensation as the critics saw it, as something akin to Fosco's almost irresistible effect on Marian's trembling body, as in the power dynamic that seduction implies. In *Crime and Punishment* Svidrigailov's attempted rape of Dunia lays bare the hierarchy of power that notions of "seduction" imperfectly mask, as well as the "cold and dead materialism" it presumes. "'Let me kiss the hem of your dress – let me, let me!' Svidrigailov says, 'I can't bear its rustling'" (*Crime* 493; *PSS* 6: 380).[12] In a Lewesian context, however, sensation is no longer a power play.

Beer writes, "[t]he need to *please* his readers as well as to unsettle and disturb them is as vital to Darwin as it was to Dickens" (Beer, *Darwin's Plots* 35). As I hope my own response suggests, pleasure is also a significant part of reading Dostoevsky. Whatever the particular sensation at hand, Dostoevsky's reliance on Lewesian "dual-aspect monism" ensures that it is always complicated, a matter of mind and body, writer's intent and reader's response, both sides of the relationship always operating as simultaneously cause and effect. It is clearly important to Dostoevsky that his readers, like his characters, come with bodies attached, bodies that engage together with our minds in the real conditions of the material world. Even so, Dostoevsky's appeal to our physical bodies in no way impinges on his trademark dialogicity but is part and parcel of the same project. Eliot, as Beer writes, came to understand her writing as a "joint enterprise of production," one that "'furnishes space' for the activity of reader and writer together'" (Beer, *George Eliot* 75–6). If the spaces in a Dostoevsky novel are more cramped all around, still our role as the reader is the same: to read Dostoevsky "for the plot" is to find our way to a truth that is also ours because we have partaken in its creation on multiple levels and with every fibre of our being. As we will see in the next chapter, just as plot indexes Dostoevsky's fidelity to Lewesian sensation, so other tropes can serve a similarly complicated materiality. These tropes include devices of a notably older, more elevated, and apparently still more strictly ornamental sort: not just plot, but metaphor and even allegory.

4 Metaphor and Allegory

As you already know, when the Underground Man says "you," he is not always addressing you, my reader. In his own historical context, the "gentlemen, sirs" that the Underground Man repeatedly invokes evidently refer to Chernyshevsky and his nihilist friends. In that case, "you" are long dead. Even should we restrict ourselves to the present moment of your reading, while you are the most immediately apparent signified to Dostoevsky's sign, you are nonetheless not the only one. When your neighbours read *Notes from Underground*, "you" refers to them, for example, whereas when I read *Notes from Underground*, "you" refers to me; "you" in English as in Russian can also mean "one," people in general, in other words, all of us taken together. In the previous chapter I suggested that the Underground Man's use of the second-person pronoun parodies the would-be intimacy that a materialism of a certain sort attempts to foist on its reader. In this chapter I would note that "you" as Dostoevsky's narrator wields it also serves a different kind of materialism altogether. As "you" elides with "me" and the Underground Man loses his way in his own shifting points of reference, his narrative is made to bear the imprint of a material reality that is itself shifting, both as shaped by its various receivers and in its own complicated and multiple existence. On this level, "you" serves not to dismantle the claims of a "vulgar" materialism, but to represent and also embody a materialism predicated instead on the mutually consti-tuting activity of minds, bodies, and world. "You" can serve these two ends at the same time because, like any pronoun, it is a sign of a certain kind: the kind that American scientist and philosopher Charles Sanders Peirce calls an index.

In his many essays published largely posthumously, Peirce presents the index first as a material trace of the sort that we already encountered in chapter two, the physical evidence that the detective reads in order to reconstruct a larger context. Peirce sees a "bowlegged man in corduroys, gaiters, and a jacket" and thinks "[t]hese are probable indications that he is a jockey or something of the sort" (Peirce, *Essential* 2: 8), just as Sherlock Holmes takes the "reddish mould" adhering to Watson's instep as a sign that his colleague has just returned from sending a telegram. For Peirce, as for the nineteenth-century detective, the reading of material traces is a scientific practice, and indices accordingly include the empirical observations that experimental scientists use to make their own kind of case, as in Louis Pasteur's "discovery" of microbes – Peirce's example as well as Latour's! – or the symptoms that reveal the presence of a disease. If Peirce himself emphasizes the means "whereby we divine the secrets of nature" (2: 224), still an index is not necessarily made of matter itself, but more broadly takes in any sign that enjoys what Peirce calls an "existential" relationship with what it represents. As a larger category, indices include deictics or shifters of all sorts, from pronouns like the Underground Man's "you" to phrases like "over there" or "next Tuesday." Here the "existential" relationship operates in terms of the real-world context that frames and gives specific meaning to each instance of the sign. As Kris Paulsen argues, the indexical quality of a sign finally is constituted not just in its relationship to its original object in the world, but also in its effect on the material being of its receiver. Here the matter of primary importance is not a set of often past material conditions, but the mind and body of a receiver who is viscerally affected by the index that she encounters.

Abduction as already noted relies on the reading of indices, with all the uncertainty that that approach entails even at its most rigorous. As Sherlock Holmes describes it in *The Hound of the Baskervilles* (1902), although "we have always some material basis on which to start our speculation," the method remains a kind of guessing; it is, as he says, "the scientific use of the imagination" (Doyle 106). The Underground Man's "you" is also an index, only especially as his own confusion underscores the awareness of contingency and different contexts that shifters make necessary. In Paulsen's argument, even plot and plot-tiness as presented in the previous chapter with their attendant "discomfort" function as indices. As Paulsen explains in language that speaks also to the workings of sensation, the index "is a phenomenal and embodied kind of sign, just as abduction is a sensuous, emotive, embodied kind of thought" (Paulsen 99). Dostoevsky's signs finally take on a strikingly indexical cast even as he turns to devices of an apparently more literary and less "existential" sort: metaphor and allegory.

The existential relationship that the index establishes is most immediately evident in more everyday uses of language as in signs that operate in the absence of words altogether, a "rap on the door," for example, or a "tremendous thunderbolt" (Peirce, *Essential* 2: 8). As Peirce himself would argue, however, to limit the indexical to what we think of as less "artistic" and more "real" contexts is to impoverish not just our art but also our science. Both sorts of limitations are readily apparent in nineteenth-century realism as it is most often defined, as a literature apparently stripped of literary device in the interest of a so-called scientific objectivity. René Wellek clearly draws on this set of assumptions when he explains that what he calls "the objective representation of contemporary social reality" relies on "the orderly world of nineteenth-century science, a world of cause and effect, a world without miracle, without transcendence even if the individual may have preserved a personal religious faith" (Wellek 241). Certainly a belief in "the orderly world of nineteenth-century science" drives Chernyshevsky's dismantling of plot, not to mention the claims of the so-called Natural School and the equally aptly named genre of the "physiology": like Wellek's, theirs is a realism defined by its apparent lack of literary artifice in an imitation of a science supposedly one with what it describes. Even in the nineteenth century, however, a scientifically inflected literature needn't take that tack.

In *Sweet Science: Romantic Materialism and the New Logics of Life*, Amanda Jo Goldstein argues that a romantic literature just a few decades earlier than Dostoevsky reached for a Lucretian materialism that claimed that facts themselves "aren't plain" (Goldstein 121). Romanticism in her reading accordingly employs a sometimes elaborate figurative language not in a departure from the real, but to better register "actual but fugitive aspects of the entities knowable through sensuous experience" (8). For this particular strain of Romanticism, she writes, "the epistemic status of poetic language *changes*: the tropes, figures, images and metaphors formerly thought to adorn or obscure the simple evidence of sense may instead index fidelity to it – may constitute the best possible linguistic register for aspects of subjective and objective activity sidelined by the philosophical pursuit of 'clear and distinct ideas'" (9). Dostoevsky's starting point is not Lucretius, but Feuerbach. He also operates in the context of a science that had advanced considerably by mid-century. His materialism is nonetheless akin to the romantic sort that Tresch also invokes, one where nature is understood not as simple and straightforward, but as "growing, complexly interdependent, and modifiable" (Tresch xi). His poetic tropes are equally far from plain.

A science predicated on the interrelationship of mind and matter lends itself not to a stripping-away of figurative language, but to its complication, first in a bidirectional metaphor that works not to elevate concrete instance over abstract referent or the other way around, but to show substance and idea as always interdependent. The mutual implication of tenor and vehicle in Dostoevsky, as in George Eliot, offers a striking indexicality of its own, as the two halves of a metaphor that point to one another act simultaneously as both cause and effect, each signifying only under the pressure of the other. The indexical quality of Dostoevsky's signs is still sharper, however, when his metaphors tip into allegory.

Allegory famously went out of fashion with a Romanticism that deemed the device, in Coleridge's words, "an abstraction from objects of the senses," a "mechanic" rather than "organic" form; in its insistent literariness and often overt didacticism, it would seem still less appropriate to the so-called realist novels that followed (Fletcher 16 fn. 29).[1] As a more recent vein of literary scholarship argues, however, far from disappearing in the wake of Romanticism, allegory proved especially useful to writers at mid-century. Dickens and Marx make pointed recourse to the device, as does Hawthorne in far-away America. In Russia, Chernyshevsky is an especially noted practitioner of the form, as *What Is to Be Done?* is awash in frequent and invariably heavy-handed allegories: the allegory of the bride, the allegory of the cellar or underground, the allegorical function of the four dreams. While any figure of speech necessarily implies two levels of meaning, Chernyshevsky's allegories are remarkable for their attempt to collapse the two into one as Chernyshevsky tries and fails to reconcile his commitment to material monism with a utopian insistence that the ideal be made real. When Dostoevsky reaches for allegory in turn, it is, as always, most obviously as a parody of the Chernyshevskyan world view. It is also, however, to reconfigure the literary device in service of a more complicated material world.

Allegory in what post-modernism, like Romanticism, has taught us to see as its open artificiality would seem entirely at odds with the "existential" relationship that indexicality entails. Certainly Dostoevsky derides the nihilist mistake of conflating reality as it is with reality as it might be, above all in *Demons*, where his mockery culminates in the "pathetic, trite, giftless, and insipid allegory" that is the quadrille of literature at the ball that is itself an allegory and that precipitates the series of calamities that concludes the novel: as the fête descends into chaos, the town goes up in flames, Shatov is killed, and Stavrogin commits suicide, Dostoevsky reveals the nihilist insistence on a single and yet ideologically correct reality as not just ludicrous but an actual dead end (*Demons* 508; *PSS* 10: 389).[2] Allegory as Dostoevsky himself practises it, however, posits the world of matter in entirely different terms. For Dostoevsky as for nineteenth-century science from Darwin, Lewes, and Maxwell to Strakhov and Peirce, materiality is not separate from our attempts to apprehend it, but instead includes our minds as well as our bodies in a "living life" that is both clearly defined and yet irreducibly multiple.

Index and Affect

Charles Sanders Peirce stands alongside his friend William James as the most significant American representative of the "alternative" nineteenth-century science that we have hitherto considered largely in its European manifestations. As Louis Menand writes in his wonderful account of this era in American thought, *The Metaphysical Club* (2001), Peirce, like many a young person in the latter half of the nineteenth century, developed his ideas first in response to Darwin. "What does it mean to say we 'know' something in a world in which things happen higgledy-pigglety?" Menand asks. "Virtually all of Charles Peirce's work – an enormous body of writing on logic, semiotics, mathematics, astronomy, metrology, physics, psychology, and philosophy, large portions of it unpublished or unfinished – was devoted to this question" (Menand 199). James Clerk Maxwell was another important influence. Like Maxwell, Menand writes, Peirce believed:

> that physical laws are not absolutely precise, and his experience as a scientist seemed to confirm this. Scientific laws rely on the assumption that like causes always produce like effects, but as Maxwell himself once put it, this assumption is a "metaphysical doctrine," one of "not much use in a world like this, in which the same antecedents never again concur, and nothing ever happens twice." (222)

"The problem," Menand explains, "boils down to the question: What does it mean to say that a statement is 'true' in a world that is always susceptible to 'a certain swerving'?" (223). For Peirce, the answer was abduction.

As compelling as its story often is, and Peirce, like many a reader of Conan Doyle, acknowledges its power, still abduction operates, as Paulsen writes, "in excess of the evidence at hand" (Paulsen 96). Abduction is the movement of the mind as, often in a flash of insight, it connects and also shapes our empirical observations. As Peirce writes in "Deduction, Induction, and Hypothesis" (1878): "I once landed at a seaport in a Turkish province; and, as I was walking up to the house which I was to visit, I met a man upon horseback, surrounded by four horsemen holding a canopy over his head. As the governor of the province was the only personage I could think of who would be so greatly honored, I inferred that this was he." "Fossils are found," he adds, "say, remains like those of fishes, but far in the interior of the country. To explain the phenomenon, we suppose the sea once washed over this land" (Peirce, *Essential* 1: 189). A science that would "solve the world's problems 'once and for all'" assumes an existential reality that we can lay bare, as we have seen in the case of Bernardian-style vivisection, even if at the expense of actual "living life." Abduction, on the other hand, knows reality only by acknowledging a fundamental and inescapable uncertainty. For Peirce, as for Strakhov, "knowledge is not a passive mirroring of the world, but an active means of making the world into the kind of world that we want it to be" (Menand 225); in Latour's words, there is no world separate from mind as itself a "wriggling and squiggling part of nature" (Latour 9).[3] With no original cause or final end, reality as abduction understands it is both real and entirely a matter of how we read its indices.

In Peirce's theory of semiotics, some signs signify by virtue of their likeness to the object that they represent, as, for example, a painting often does. Others are a matter of convention; as Paulsen explains, "A red traffic light indicates 'stop,' and a green, 'go,' simply by 'consequence of habit'" (Paulsen 86). Peirce defines the index in contrast in terms of its "existential" relationship to what it represents, first as "a real thing or fact which is a sign by virtue of being connected with it as a matter of fact." (Peirce, *Collected* 4: 359). A "real thing" might be the material trace of a past event, as in "a piece of mould with a bullet-hole in it as a sign of a shot" (Peirce, *Philosophical* 104). It might also mark a present context, as Peirce explains:

> a low barometer with moist air is an *indication* of rain; that is, we suppose that the forces of nature establish a probable connection between the low barometer with moist air and coming rain. A weathercock *indicates* the direction of the wind…. The pole star is an *index*, or pointing finger, to show us which way is north. (Peirce, *Essential* 2: 14)

Indexicality doesn't necessarily lie in the materiality of the sign itself, however. As Peirce writes, while a map is always a likeness, "unless it carries a mark of a known locality, and the scale of miles, and the points of the compass, it no more shows where a place is than the map in *Gulliver's Travels* shows the location of Brobdingnag." The difference that indexicality makes is not a change in the material status of the map itself; even as a likeness, a map is an object in the world. It is instead our "experience of the world we live in ... [that] ... renders the map something more than a mere *icon* and confers upon it the added characters of an *index*" (2: 8). Like a map, but in the absence of any material reality of their own, words like "you" or "there" or "on the right (or left) of" (Peirce, *Philosophical* 111) also signify only when they point to a particular instance in the world. Deictics, or words that point, insist on this particularity even when they evade it, as the fraught example of *Notes from Underground* especially well illustrates.

A large part of the difficulty that *Notes from Underground* presents derives from our inability to orient ourselves in the text. It's not that the Underground Man's pronouns lack the concrete referents that would give them meaning, but that they offer too many. While the text begins, "I am a sick man" (*Notes* 3: *PSS* 5: 99), a note at the bottom of the page immediately forestalls any attempt to associate the first-person pronoun with Dostoevsky himself. The note begins: "*Both the author of the notes and the *Notes* themselves are of course fictional." That the author is fictional, however, doesn't mean that he's not real. As the note continues, "Nevertheless, such persons as the writer of such notes not only may but even must exist in our society, taking into consideration the circumstances under which our society has generally been formed" (3; *PSS* 5: 99). "I" in this sense is all or at least some of us, as the Underground Man makes clear when he turns to "we" in the final pages of his narrative: "we've all grown unaccustomed to life, we're all lame, each of us more or less" (129; *PSS* 5: 178). In the radical shifting noted at the end of the previous chapter, "you" in turn points not just to Chernyshevsky et al., to you and to me, but sometimes even to "me," as in the Underground Man himself.

"You" in the latter case appears in quotes when the narrator attempts to anticipate his reader's response. "'But is this not shameful, is it not humiliating!' you will perhaps say to me, contemptuously shaking your heads. 'You thirst for life, yet you yourself resolve life's questions with a logical tangle. And how importunate, how impudent your escapades, yet at the same time how frightened you are!'" (*Notes* 38; *PSS* 5: 121). As the Underground Man works to shift responsibility for his own unpleasant behaviour, however, the "you" that is really "me" is also given in his own voice. I would maintain, for example, that it is the Underground Man who wants to talk about slaps, not me, this despite his disclaimer: "But, enough, not another word on this subject which you find so extremely interesting" (12; *PSS* 5: 105). Even as our narrator claims to prize above all "our personality and our individuality" (28–9; *PSS* 5: 115), his attempts to anticipate our response and to replace it with his own undermine the existential relationship that indices need in order to signify. We are left in his own account with an unnamed narrator and no reader at all. "And here's another puzzle for me," he writes, "why indeed do I call you 'gentlemen,' why do I address you as if you were actually my readers? Such confessions as I intend to begin setting forth here are not published and given to others to read" (39; *PSS* 5: 122). The contradiction of his self-declared missing "existential" relationship, however, is exactly Dostoevsky's point.

If the Underground Man makes for difficult reading, from a Peircean point of view, "you" as Chernyshevsky refuses to problematize it is unnaturally fixed, and the Underground Man in his insistence on shifters that keep shifting lays bare the abstraction that Chernyshevskyan-style materialism in fact offers. Although a real other nonetheless eludes the Underground Man, with his vivid awareness of contingency and context, he also only just misses the mark. In an essay published shortly before *Notes from Underground*, "Winter Notes on Summer Impressions" ("Зимние заметки о летних впечатлениях," 1862–3), Dostoevsky responds to what he sees as the social fragmentation so characteristic of the West. As Konstantin Mochulsky summarizes his argument, whether Western society forms itself along traditionally capitalist or newly socialist lines, either way "the personal principle, the 'self-determination in one's own I,' … hinder[s] the formation of brotherhood" (Mochulsky 234). The answer that Dostoevsky proposes for Russia, however, is not to abandon "the personal principle," but to re-cast it. "Understand me," he writes, "voluntary, fully conscious self-sacrifice, free of any outside constraint, of one's entire self for the benefit of all is, in my opinion, a mark of the highest development of individuality" (Mochulsky 234). We rightly see in this quote an expression of Dostoevsky's Christianity, and in "Winter Notes" he even cites the example of Christ on the cross. The permeability of self and other, mind and material world, is also, however, a fundamental tenet of materialism as both Peirce and also Lewes understand it. In *Notes from Underground*, this idea is realized on both levels – as science as well as faith – in Liza.

It is the Underground Man's encounter with Liza that motivates the entire narrative, as he acknowledges at the end of Part One when he proposes that he tell "this anecdote that now refuses to be gotten rid of" (*Notes* 41; *PSS* 5: 123). The climax of "Apropos of Wet Snow" ("По поводу мокрого снега") in turn is the "irrepressible impulse" (123; *PSS* 5: 175) that prompts Liza to reach past the Underground Man's angry words and embrace him. As the Underground Man explains, it is only at this moment that he sees Liza as a person in her own right for the first time. "I was so used to thinking and imagining everything from books," he writes, "and to picturing everything in the world to myself as I had devised it beforehand in my dreams, that at first I didn't even under-stand this strange circumstance" (123; *PSS* 5: 174). "What occurred," he explains, "was this: Liza, whom I had insulted and crushed, understood far more than I imagined. She understood from it all what a woman, if she loves sincerely, always understands before anything else – namely, that I myself was unhappy" (123; *PSS* 5: 174). Unfortunately, the Under-ground Man himself is capable only of flashes of empathy with a real other, and when he immediately attempts to return Liza to the role of paid prostitute that he has already imagined for her, he rejects the con-nection with a real person in the world that her freely given embrace offers. Even as the Underground Man in his own fictional context backs away from the shifting realities that "you" and "I" indicate, however, the discomfort that he produces not in her but in us presents an indexi-cality of yet another sort.

The Underground Man is not a sensationalist in the way that Dosto-evsky himself so often is. While his narrative features prostitution and even a certain amount of drunkenness, still it offers nothing like "the themes of inheritance, bigamy, poisoning, drug abuse, and adultery, and … frequent employment of the *deus ex machina* and other startlingly improbable coincidences" (Fantina 23) that characterize *The Idiot* or *The Adolescent* (*Подросток*, 1875) as they do the British novel of sensation. Even so, the Underground Man has his own means of "titillating" and "tormenting" his reader, not just his troubling use of pronouns, but his open abuse of the reader, including his painfully detailed renderings of his own humiliation. On all counts, *Notes from Underground* offers an acutely uncomfortable reading experience. As the Underground Man describes his own act of writing as "no longer literature, but corrective punishment," so, too, our act of reading; in the words of my student once more, "Painful book to read. Cringing." In "The Index and the Interface" (2013) Paulsen focuses on affect as a measure of indexicality in order to make the claim that photographs remain indexical even in a digital age. Her argument applies equally well, however, to the "hor-ripilation" that sensation intends.

Paulsen's article hinges on an art historical tendency to read Peirce a little selectively. In a particularly famous example in "What Is A Sign?," Peirce describes the still new technology of analogue photography as a combination of two sorts of signs, the icon, or likeness, and the index. "Photographs," he explains:

especially instantaneous photographs, are very instructive, because we know that they are in certain respects exactly like the objects they represent. But this resemblance is due to the photographs having been produced under such circumstances that they were physically forced to correspond point by point to nature. In that aspect, then, they belong to the second class of signs, those by physical connection. (Peirce, *Essential* 5–6)

In the twenty-first century, digital technology has cast that "physical connection" in doubt, and Paulsen writes to restore indexicality to photographs of all kinds. She first notes the mistake that we make when we assume that we have somehow lost materiality when images are no longer made "by a physical touch or imprint, but by the numerical sensors that translate light into data" (Paulsen 85). More importantly, she argues for a more complete reading of Peirce. In Peirce's own account, an index signifies not just as "a real thing or fact which is a sign by virtue of being connected with it as a matter of fact," but also by "forcibly intruding upon the mind, quite regardless of its being interpreted as a sign" (Paulsen fn 5, 106; Coll Pap ¾ 4: 359). In the latter instance, the index creates meaning not as a material trace, but in its material effects on the receiver.

"Anything which focuses the attention is an index," Peirce writes, "Anything which startles us is an index, in so far as it marks the junction between two portions of experience" (Peirce, *Philosophical* 108–9). A rap on a door operates in these terms, as do the demonstrative pronouns "this" and "that," "[f]or they call upon the hearer to use his powers of observation, and so establish a real connection between his mind and the object" (110). Peirce's best and most complete example is the word "Hi!":

> When a driver, to attract the attention of a foot-passenger and cause him to save himself, calls out 'Hi!', so far as this is a significant word, it is, as will be seen below, something more than an index; but so far as it is simply intended to act upon the hearer's nervous system and to rouse him to get out of the way, it is an index, because it is meant to put him in a real connection with the object, which is his situation relative to the approaching horse. (Peirce, *Essential* 2: 14)

Paulsen understands the index as affect also in terms of what Roland Barthes calls the *"punctum."* In *Camera Lucida* (1980) Barthes describes the *"punctum"* as that "element" of a photograph that "rises from the scene, shoots out of it like an arrow, and pierces me" (Barthes 26). Barthes's *punctum*, Paulsen argues, adds indexicality to a photograph that would otherwise remain an icon alone, as it "causes a sensory encounter (from ground to the sign to the receiver), which then starts a chain of associative and exploratory thoughts." Like an index, Paulsen writes, the *punctum* "'expands' and points to something 'beyond' what is merely represented" (Paulsen 93). We might note that indexicality in this sense distinguishes all art from signs of a more everyday sort. Even within the already affective realm of art, however, some signs are more pointedly indexical than others.

A dialogue between a constantly shifting "I" and "you" forces consideration of our "existential" reality even as it renders that reality highly uncertain. By the same token, the Underground Man, like the nineteenth-century novel of sensation and like the word "Hi!" in Peirce's imaginary scenario, relies for his effects on the only partly predictable response of our own bodies. Either way, Dostoevsky indexes his fidelity to a "sensuous, emotive, embodied kind of thought" (Paulsen 99) and a world "susceptible to 'a certain swerving'" (Menand 223). Like the effects of plot as the nineteenth century often saw it, the narrator in *Notes from Underground* is nonetheless somewhat of a blunt instrument, and his tack only barely literary. Indexicality attaches to Dostoevsky's signs even when those signs are of a less everyday and apparently more conventionally artistic sort, however, although its workings are then a little less "plain." In the case of the complicated metaphoricity that Dostoevsky shares with George Eliot, indexicality is no longer a matter of clue, deictic, or even affect. In Paulsen's argument, the *punctum* as it "reaches out and touches the viewer, pierces her, bruises her, and then causes her to reach back and explore its root and its cause" engages both viewer and photographic image in a "mutual and reflexive" relationship (Paulsen 93). As tenor and vehicle in Dostoevsky and Eliot point to and shape one another, "mutuality and reflexivity" mark the very operation of their figurative language. Their bidirectional metaphors also point out of the text and to a particular feature of existential reality as Peirce also knows it: Lewesian dual-aspect monism.

Eliot and Dostoevsky, II: Metaphor

Against the background of a more sober and stripped-down realism, Dostoevsky is striking in his overt and extended figuration. Dostoevsky's characters tend to come freighted with multiple significations, so Prince Myshkin in *The Idiot*, for example, is not just an idiot in various meanings of the word as well as a failed Westernized liberal, but also Christ at the Second Coming, while Nastasya Filippovna is simultaneously a real victim of rape, a living embodiment of the "Woman

Question," and an image of fallen Beauty. As Konstantin Mochulsky elaborates on the latter level of meaning:

> *On the metaphysical plane* ... [Dostoevsky's] ... heroine is the image of "pure beauty" seduced by the "prince of this world" and waiting in her dungeon for a liberator. The soul of the world, the beautiful Psyche – existing in the bosom of the divinity, on the boundary of time – fell away from God.... From her former existence outside of time, Psyche has preserved memories of the 'sounds of heaven' and a feeling of fatal, irreparable guilt. The evil spirit that seduced her excited pride and a consciousness of guilt in the exile and through this drives her to destruction.
>
> And here a man comes to her with tidings about her heavenly homeland. He too has been in the "shade of the gardens of paradise", he saw her there in the "image of pure beauty" and in spite of her earthly degradation, recognizes his other-worldly friend.... Thus takes place the mystical meeting of the two exiles from paradise. Vaguely they remember their heavenly homeland ... as "in a dream." (Mochulsky 377–8)

In similarly involved fashion, brothers in *The Brothers Karamazov* refer simultaneously to the four (half) brothers, the monks at the nearby monastery, and the brotherhood of man, while the (dead) fathers are Fyodor Pavlovich, Father Zosima, and God the Father as Ivan in his unbelief would also have him. This often stunning degree of over-determination reaches a height in the simultaneously real and symbolic meanings that Dostoevsky attaches to space.

In *Notes from Underground*, the "underground" [подполье] that Dostoevsky erects in direct response to Chernyshevsky's "cellar" [подвал] renders a philosophical stance in material terms, as the walls that demarcate the unnamed narrator's living space explicitly stand also for the limitations of "vulgar" science. "What stone wall?" the Underground Man asks. "Well, of course, the laws of nature, the conclusions of natural science, mathematics" (*Notes* 13; *PSS* 5: 105). On one level the walls of his shabby apartment frame the narrator's encounters with Liza and with his servant Apollon as they shield him from the cold St. Petersburg climate. As the "laws" that insist on a strict separation of self and other, however, the walls of the underground also separate the Underground Man from Liza, from "living life," and ultimately from his own self: as long as he separates himself from the material world that includes both our minds and our bodies, even in his own fictional context, our unnamed narrator has no reality of his own. This paradoxical combination of an insistent materiality and multiply symbolic meaning is still more striking in the setting of *Crime and Punishment*.

In *Crime and Punishment*, the city of St. Petersburg is represented with such exactness that we can trace the characters' movements on a map; the real, material city is also emphasized in the many references to the heat, the smell, and the lack of air. At the same time, the novel very evidently takes place also in Raskolnikov's own mind, as the cluttered physical space, cluttered not just with things, but with characters who reflect aspects of our hero back to himself, is equally a representation of Raskolnikov's cramped mental confines. What Raskolnikov needs, as Porfiry Petrovich tells him, is "air, air!," both the fresh air that the city so pointedly lacks and a spiritual way out that is materialized in turn in a choice between two real geographical locations, America and Siberia, that again enjoy a symbolic dimension: Siberia is life and America a symbolic death made real, as, where Chernyshevsky's Lopukhov only pretends to kill himself but really emigrates to America, Dostoevsky's Svidrigailov announces his departure for America only to commit suicide.[4]

Central to this multifaceted rendering of space is the historical reality of St. Petersburg itself, a real locale that was always also a symbol, Peter the Great's "window on the West" or, in the Underground Man's words, not just the most "intentional" but the most "abstract" city on earth. Equally important is Dostoevsky's adherence to a Christian faith that always operates on multiple planes. As Tatiana Kasatkina explains:

> under the drunken, pothouse palpability of what Dostoevsky called "the most intentional city on earth" are glimpsed the imperishable outlines of the true reality: Dunya is a third-century martyr or fourth-century desert-dwelling anchoress; Raskolnikov an ascetic in the Egyptian desert, who has been redeemed and now needs to commit a new crime so as to die … and Sonya lives in Capernaum …, the city where Christ and his disciples found refuge during His three-year ministry and which was the home town of Peter and Andrew, the first to be called. (Kasatkina 32)

Dostoevsky's complicated conflation of physical and mental or spiritual spaces finally also points to a specific aspect of existential reality, as Lewes described it. Dual-aspect monism, as it turns out, partakes of the indexical, or, more likely, the other way around.

As Tresch notes, well before Peirce claimed "a low barometer with moist air" as an "*indication* of rain," barometers already served as a species of romantic machine.[5] Again before Peirce, the device also appears in *Problems of Life and Mind*, although now not as an actual object in the world. For Lewes, as for Peirce, the barometer serves instead as an icon, or likeness, of a mind whose signs are indexical in turn. Lewes in *Problems of Life and Mind* reaches for an imaginary barometer as he carves out a stance neither entirely realist nor entirely idealist. Is the world as we know it "figured in Feeling through the *adaptation* of the sentient organism to the external Real," he asks, "or is it simply a *subjective construction* – the figuration of Sensibility – which we illusively project outwards and receive back again in reflection?" (Lewes, *Problems* 1: 170). His answer to his own either/or question is a qualified "yes." As he explains with recourse to physiology, "the pulpy mass of the brain acquires, through manifold experiences, a structure more and more variously definite, with corresponding reactions; and as Feeling becomes differentiated and defined, Qualities arise in the Felt" (1: 169). As he then puts it in the parallel philosophical terms that we already know from Franklin Blake: "the object is necessarily object-subject, and subject is equally subject-object" (1: 171). Lewes is careful to clarify his meaning:

> I do not agree with those realists who conceive the thing represented in Perception, in the way mathematicians regard an algebraic function as represented by a curve, – object and subject forming a Dualism having something of a pre-established harmony, but no real union. I would rather liken the Thing represented in Perception, to the weight of the atmosphere represented by the height of the mercury in the barometer; while the differences between weight and height, and between atmosphere and mercury, are wide, both rest on a common identity of pressure. (1: 171)

As Peirce recognizes in his own turn to semiotics, science, all science, is a science of signs. Even so, Lewesian dual-aspect monism deals in signs of a particularly Peircean sort. If, as Lewes writes, the "Thing represented in Perception" is a sign, it is a sign joined to the Thing that it signifies by a "common identity of pressure" that the two halves of this relationship, mind and matter, exert on one another, to borrow now from Peirce, each as a "a real thing or fact which is a sign by virtue of being connected with it as a matter of fact." Like Peirce's, the indexicality that Lewesian science entails is finally more than a matter of content. It is also a rhetorical strategy.

Because science itself is told in words, even scientists of a "vulgar" sort speak in figures. We might recall Bernard with his comparison of science to a "long and ghastly kitchen," for example, or Moleschott's personification of "Poor Ireland!" A reliance on rhetorical device is nonetheless much more marked in a science predicated on the mutual and reflexive relationship of mind and matter. Maxwell, for one, openly relies on simile, or what he calls "physical analogy." "By a physical analogy," Maxwell explains, "I mean that partial similarity between the laws of one science and those of another which makes each of them illustrate the other" (Maxwell 156). Darwin's device of choice, on the other hand, is metaphor, above all in the first edition of *On the Origin of Species*. As Beer argues, in subsequent editions Darwin defended his theory "by paring away multiple significations, trying at points of difficulty to make his key terms mean one thing and one thing only" (Beer, *Darwin's Plots* 34), a belated stripping-away that came at a cost to an argument premised on multivalence and "superfecundity" (6). Darwin's "exuberantly metaphorical drive," Beer writes, was "proper" (34) to his theory as he originally conceived it, as "[t]he unused, or uncontrolled, elements in metaphors such as 'the struggle for existence' t[ook] on a life of their own" (6). In Latour's account, the signs of a multivalent and uncertain science acquire an explicitly indexical cast.

While Latour writes to claim an entirely new way of doing science, he nonetheless gives a nod to William James when he casts aside mimesis as traditionally conceived for what he calls a "much more reliable movement – indirect, cross-wise, and crablike – through successive layers of transformations." Just as Lewes looks to a "common identity of pressure," so Latour argues that "reference is not simply the act of pointing or a way of keeping, on the outside, some material guarantee for the truth of a statement; rather it is our way of keeping something *constant* through a series of transformations." While this model, like abduction, "forfeit[s] resemblance," it claims its own compensations. "[B]y pointing with our index fingers to features of an entry printed in an atlas," Latour explains, "we can, through a series of uniformly discontinuous transformations, link ourselves to Boa Vista" (79) or, in other words, connect ourselves to a real place in the world. As it spills into literature, this science and the indexicality that it entails take shape in a set of figures that we can only call realist.

As Roman Jakobson noted long ago, the word "realism" all on its own is nothing more than a commitment to "conveying reality as closely as possible" (20) that functions in different ways in different eras and across different genres of art. The nineteenth-century literary movement that bears the same name, however, has consistently been understood as implying a scientific world view. This point holds even when that science takes the form of Comte's "social physics." To quote Wellek again, what he calls "the objective representation of contemporary social reality" relies on "the orderly world of nineteenth-century science, a world of cause and effect, a world without miracle, without transcendence even if the individual may have preserved a personal religious faith" (Wellek 241). Nineteenth-century realists often make their scientific underpinnings clear, when Balzac frames *The Human Comedy* (*La Comédie humaine*, 1842) with reference to zoology, for example, or when Zola claims the mantle of Bernard in his "experimental" novel, or when Chernyshevsky writes of "artificial albumin" and four cups of coffee. When scientific facts are no longer "plain," however, the realist literature that would tell them isn't "plain," either, not just in Dostoevsky, but even in Tolstoy.

To the extent that he attempts in *War and Peace* to do without the bells and whistles of plot, Tolstoy presents as a realist as traditionally conceived, the kind who would do away with literary artifice altogether. As his embrace of calculus argues for a more complicated reality, however, so does his (qualified) use of figurative language. A figure, any figure, operates on two levels, one as if it were "real," and the other present only by analogy, and *War and Peace* is remarkable for its extraordinarily long and detailed similes, the abandoned Moscow as a queen-less beehive across two pages of text, for example, or, on two different occasions, Prince Andrei as the old oak tree. The care that Tolstoy takes to spell out the terms of his comparisons – "'Yes, it's right, a thousand times right, this oak,' thought Prince Andrei, 'Let others, the young ones, succumb afresh to this deception, but we know life – our life is over!'" (L. Tolstoy 420) – speaks to his ever-present desire to flatten the two levels of his language into one historical truth. As these same signs shift under the pressure of the very lives that they indicate, however, Tolstoy, as always as if against his will, acknowledges an essential multiplicity.

Even when true and expressed with maximum clarity, meaning in Tolstoy always assumes a mind that is itself matter and a material world that is only what we know it to be. The same tree speaks different words to Prince Andrei at two different points in both their lives, for example, just as the comet that tells Pierre of his love for Natasha presages for others "all sorts of horrors and the end of the world" (L. Tolstoy 600). The ambivalent workings of Lewes's "common identity of pressure" are especially apparent in the recurring device of Pierre's dreams. While Pierre's dreams at Mozhaisk and again at Shamshevo illuminate for Pierre as well as for the reader the true meaning of Pierre's life, they are also very evidently shaped by the circumstances of that life in a circularity that undermines any simple notion of objectivity. Just as the sleeping Pierre decides that his aim must be "to *hitch together* all these thoughts – that's what's needed! Yes, *we must hitch together, hitch together!*,'" he awakens to the voice of his groom: "'We must hitch up, it's time to hitch up, Your Excellency! Your Excellency! … we must hitch up, it's time to hitch up'" (844). If the message of Pierre's dream is no less true for its contingent and circumstantial nature, still Tolstoy is evidently at pains to keep his figurative language in check. We find a freer hand with rhetorical flourish in a realist writer whose Lewesian underpinnings are, as always, more immediately evident: George Eliot.

Like Tolstoy, Eliot is not particularly well known for her recourse to the further reaches of literary art. Where Henry James saw Eliot's achievement in *Middlemarch* as diminished by a style that he dismissed as "too often an echo of Messrs. Darwin and Huxley," for Virginia Woolf, she is just a little dull, this apparently in direct contrast to Dostoevsky. In her *Common Reader* (1925), Woolf's essay "George Eliot" is followed immediately by "The Russian Point of View." In the latter, Woolf's Dostoevsky is marked by "seething whirlpools, gyrating sandstorms, waterspouts which hiss and boil and suck us in" (Woolf 250). In the former, her Eliot is very good and more than a little ponderous, with "none of that romantic intensity which is connected with a sense of one's own individuality, unsated and unsubdued, cutting its shape sharply upon the background of the world" (234–5). Certainly, the difference in pace and tone is dramatic, and we can't help but note the relative absence of drunkenness, dirt, and attacks of brain fever in Eliot's Middlemarch as opposed to Dostoevsky's St. Petersburg. Even so, readers of the Woolfian sort overlook the real weirdness of Eliot's images.

As James's criticism at least half-acknowledges, Eliot's dips into the language of science are often startling in their power, as when Dorothea in Rome discovers "red drapery which was being hung for Christmas spreading itself everywhere like a disease of the retina" (Eliot, *Middlemarch* 194). Striking, too, is Eliot's occasional use of a very different register, for example in the narrator's deeply strange explanation of why the still-married Dorothea longed to see Will. "How could it be otherwise?" the narrator asks: "If a princess in the days of enchantment had seen a four-footed creature from among those which live in herds come to her once and again with a human gaze which rested upon her with choice and beseeching, what would she think of in her journeying, what would she look for when the herds passed her?" (539). What I can only call the "romantic intensity" of Eliot's figurative language is finally on full display in a representation of conjoined physical and mental or spiritual space that rivals Dostoevsky's for complexity and force.

As Kate Flint argues, Eliot's overlapping of physical and mental space is most often associated with Dorothea, for example, when Dorothea moves to Lowick only to find "that the large vistas and wide fresh air which she had dreamed of finding in her husband's mind were replaced by ante-rooms and winding passages which seemed to lead nowhither" (Eliot, *Middlemarch* 195). It is also notable in the case of Lydgate. Lydgate, we learn, discovered medicine as a boy when a wet day sent him to the encyclopedia for entertainment and the entry on Anatomy introduced him to the valves of the heart. As the narrator explains: "He was not much acquainted with valves of any sort, but he knew that *valvae* were folding doors, and through this crevice came a sudden light startling him with his first vivid notion of finely adjusted mechanism in the human frame." At that moment, "the world was made new to him by a presentiment of endless processes filling the vast spaces planked out of his sight by that wordy ignorance which he had supposed to be knowledge." It is Lydgate's tragedy that he eventually succeeds in stepping into these "vast spaces" only to find them transformed. As the narrator already warns with recourse again to spatial metaphor:

> In the multitude of middle-aged men who go about their vocations in a daily course determined for them much in the same way as the tie of their cravats, there is always a good number who once meant to shape their own deeds and alter the world a little. The story of their coming to be shapen after the average and fit to be packed by the gross, is hardly ever told even in their consciousness; for perhaps their ardour in generous unpaid toil cooled as imperceptibly as the ardour of other youthful loves, till one day their earlier self walked like a ghost in its old home and made the new furniture ghastly. (144–5)

Indeed, ghastly new furniture is exactly Lydgate's fate.

The proximate cause for the collapse of Lydgate's marriage with Rosamond is the furniture she wants that they can't afford. The furniture is also mental, however, as the narrator explains that Lydgate:

> was beginning now to imagine how two creatures who loved each other, and had a stock of thoughts in common, might laugh over their shabby furniture, and their calculations how far they could afford butter and eggs. But the glimpse of that poetry seemed as far off from him as the carelessness of the golden age; in poor Rosamond's mind there was not room enough for luxuries to look small in. (Eliot, *Middlemarch* 701)

The stakes are higher in *Crime and Punishment* and the spaces more cramped on every level, but Rosamond is trapped within and without just like Raskolnikov in his "cupboard" of a room. "What an awful apartment you have, Rodya; like a coffin," his mother says. And later:

"his room is awfully stuffy … but where can one get any air here? It's the same outside as in a closed room. Lord, what a city!" (*Crime* 231, 241; *PSS* 6: 178, 185). For Dostoevsky's characters, as for Eliot's, inside and outside are one, in Lewesian terms, "the two being as the convex and concave surfaces of the same sphere." The simultaneously real and metaphorical spaces that mark both their novels operate in the same dual-monistic fashion.

Metaphor, like simile, is premised on a notion of two halves, tenor and vehicle, or ground and figure, or target and source, one an item at hand and the other present only as a point of comparison. As George Lakoff and Mark Johnson argue in *Metaphors We Live By* (1980), we then understand the movement between those two halves as one-way. In their words, "there is *directionality* in metaphor, that is, that we understand one concept in terms of another" (Lakoff and Johnson 112). In Dostoevsky and Eliot, however, the "directionality" goes both ways. Especially as it is often the substance of the analogy that is emphasized, valves that become more real as the doors opening on to a new world view, for example, or an imaginary room that partakes of more reality or at least more materiality than Rosamond's real mind, the representation of space in both Dostoevsky and Eliot works not to elevate concrete instance over abstract referent or the other way around, but to show substance and idea as always interdependent: just as mind and body shape and imply one another, so Casaubon and his house, or Raskolnikov and his city, point to one another. As we oscillate between two halves of a figure both equally real and equally symbolic, Dostoevsky's and Eliot's metaphors accordingly both represent *and* embody (the two actions being in some respects the same) the interrelationship of mind and matter that is Lewes's claim.

Dostoevsky's metaphors, like Eliot's, offer a remarkably beautiful and also complete expression of the mutuality and reflexivity that marks this other sort of science. Even granting that achievement, still Dostoevsky's indexicality is at its most striking when he stretches his realism just a little further to include yet another rhetorical device, one apparently still more distant from "existential" reality: allegory. Despite her reputation, allegory isn't entirely lacking in Eliot. An especially notable example occurs late in the novel when Dorothea, up early to offer her assistance to Rosamund, glimpses "a man with a bundle on his back and a woman carrying her baby" (Eliot, *Middlemarch* 788), a shepherd in the distance. Although Eliot is less than explicit, still it's difficult to see her signs here as pointing other than to one family in particular: the Holy Family.[6] We might also note the revelation at the novel's very end that Dorothea Brooke's name has been allegorical all along. As the final lines in this long novel explain:

> Her finely touched spirit still had its fine issues, though they were not widely visible. Her full nature, like that river of which Cyrus broke the strength, spent itself in channels which had no great name on the earth. But the effect of her being on those around her was incalculably diffusive: for the growing good of the world is partly dependent on unhistorical acts; and that things are no so ill with you and me as they might have been, is half owing to the number who lived faithfully a hidden life, and rest in unvisited tombs. (838)

Even so, and as we might expect for a writer of what are again conventionally termed realist novels, Eliot wields allegory only on occasion and with the same degree of understatement that marks her recourse to plot. Dostoevsky, on the other hand, knows no such restraint.

Indeed, as Woolf would agree, the great pleasure of reading Dostoevsky is that he never does know restraint. Just as Dostoevsky's "sensational" plots are replete with adultery, murder, and "other startlingly improbable coincidences," so allegory in his novels runs rampant, so much so that the characters themselves recur to the very terms "allegory" and "allegorical" with remarkable frequency. In *The Idiot*, Lebedev as a self-styled "professor of the Antichrist" describes his interpretation of the Book of Revelation as his "unrolling" of "the allegorical scroll" (*Idiot* 201; *PSS* 8: 168), while Versilov in *The Adolescent* smashes an icon and shouts, "Don't take it as an allegory!" only to add with utter inconsistency: "But, anyhow, why not take it as an allegory; it certainly must have been" (*Adolescent* 508; *PSS* 13: 409). This tendency reaches its height in *Demons*, where the characters' often absurd allegorical assertion of matter as the only measure of reality serves as a direct mockery of the "vulgar" world view. Even as he mocks nihilist would-be monism, however, Dostoevsky himself holds to an allegory of a stunningly dual-monist, indexical sort, one that claims clearly defined yet multiple meanings not as a departure from "existential" reality, but as its more adequate expression.

Allegory and Index

Allegory, in its open artificiality, would seem so far from "plain" as to be entirely at odds with the representation of reality that science intends. With significant exceptions – *The Chronicles of Narnia* come to mind – the device also went out of literary fashion with the advent of Romanticism.

It was first Goethe who distinguished allegory "where the particular serves only as an example of the general," from the truly poetic device of symbol, "where the particular represents the more general, not as a dream or a shadow, but as a living momentary revelation of the Inscrutable" (Fletcher 13, fn 24). Coleridge then made Goethe's claim into a distinction between what he termed "mechanic" and "organic" form. Allegory for Coleridge "is but a translation of abstract notions into a picture-language, which is itself nothing but an abstraction from objects of the senses." Symbol, on the other hand, "always partakes of the reality which it renders intelligible; and while it enunciates the whole, abides itself as a living part in that Unity of which it is the representative" (Fletcher 16, fn 29). While the romantic call for poetic immediacy and a kind of naturalness has apparently permanently relegated allegory to the literary margins, still Romanticism didn't quite succeed in doing away with the device altogether. Instead, allegory makes a quick if often unrecognized comeback.

Allegory as writing with what Dostoevsky would call a "tendency" is especially prominent in Chernyshevsky's unabashedly ideological *What Is to Be Done?* When Marya Alekseevna drops hints about his supposed fiancée, Lopukhov asks himself, "Why did I devise such an allegory – it wasn't needed at all!" (Chernyshevsky, *What?* 106). If it wasn't needed, the allegory pointedly continues, first as Vera Pavlovna enters into a fictional marriage with Lopukhov and then as the "bride of her bridegroom" in Vera Pavlovna's fourth dream represents the future of Vera Pavlovna herself. The didacticism that marks allegory as opposed to other, more open forms of figuration already insists on a singular meaning. As Angus Fletcher explains, since allegory "implies a dominance of theme over action and image, … the mode necessarily exerts a high degree of control over the way any reader must approach any given work" (Fletcher 304). Chernyshevsky makes very sure of that control, however, through the simple device of repetition.

In her first dream, Vera Pavlovna sees herself "locked up in a damp, dark cellar [подвал]" when the door "suddenly" flies open and "she finds herself in a field, running about and skipping" (Chernyshevsky, *What* 129–30). After she recounts her dream to Lopukhov, it promptly comes true, as she says, "So, my dear, you are liberating me from this cellar," and she then makes the same allegorical reference again and again: "I now know that I'm leaving this cellar"; "I shall escape from this cellar!"; "How did I manage to breathe in that cellar?" until the allegory migrates to the narrator who addresses us, his readers: "Come up out of your godforsaken underworld [из вашей трущобы], my friends" (Chernyshevsky, *What* 143, 149, 151, 179, 313). As Chernyshevsky forcefully urges his readers to flatten the two halves of his allegory into one reality, the connection with his social-utopian aspirations is clear: allegory in Chernyshevsky allegorizes what the novel presents as the function of art more generally, which is to make real its own fictions. This movement from text straight out to world offers a curious inversion of Chernyshevsky's claim in his dissertation, *The Aesthetic Relations of Art to Reality* (Эстетические отношения искусства к действитеьности, 1855), that the function of art is not to create but to "reproduce nature and life" (Chernyshevskii, *Estetika* 154). Either way, however, "nature and life," the real or material world, stands above and apart from our attempts to apprehend it, as Chernyshevskyan materialism paradoxically insists on the very duality that it denies.

Like Quetelet when he replaces the particularity of individual and living human beings with the abstraction of an *"homme moyen,"* so Chernyshevsky's allegorical insistence that the ideal be made real renders abstract both halves of his figure. Now the item at hand is an obvious artifice, while the analogy stands as the supposed "real" reality that in fact exists only in the shadowland of the much-desired future. As the last chapter has already argued, the result is exactly as Coleridge warns: as Chernyshevsky's ideological insistence empties his materialism of any real matter at all, his allegory entirely fails to "partake[] of the reality which it renders intelligible." As the examples of Goethe and Coleridge illustrate, the romantic response to the Enlightenment as an earlier iteration of Chernyshevsky's combined rationalist *and* empiricist project was to reject allegory altogether in favour of what Murray Krieger calls "a form-making power that could break through the temporal separateness among entities, concepts, and words to convert the parade of absences into miracles of co-presence" (Krieger 4). More recently, Paul de Man simply accepts the belatedness of language. As Krieger argues, the post-modern attempt to recuperate allegory returns us to the same "vulgar" belief in a "bedrock existential reality," only now combined with an embrace of the inevitable non-coincidence of sign and signified that a "bedrock" reality makes necessary. For de Man, the great virtue of allegory is not its drive to make dreams real, but its open acknowledgment of "the fallen world of our facticity" (Krieger 16). If we cast mind as part of a material world that is itself multiple and even shifting, however, allegory can serve other ends.

Karl Marx is another notable allegorist at mid-century, if largely for the same nihilist reasons: even while his friend Engels was at pains to distinguish dialectical materialism from materialism of the "vulgar, itinerant-preacher" sort, Marx's own attempt to replace everyday objects with the "real" reality of economic relationships suffers from the same abstraction. The allegories that ensue are nonetheless a good deal weirder than anything we find in Chernyshevsky, and Theresa Kelley gives the example of his extended representation of the transformation from wood to table and table to commodity. According to Marx:

> The form of wood, for instance, is altered if a table is made out of it. Nevertheless, the table continues to be wood, an ordinary, sensuous thing. But as soon as it makes its entrance as a commodity, it changes into a thing which transcends sensuousness. It not only stands with its feet on the ground, but, in relation to all commodities, it stands on its head, and evolves out of its wooden head grotesque ideas, far stranger than if it were to begin dancing of its own free will. (Kelley 217)

As Kelley argues, "The table-become-commodity reproduces what commodification does to things: it turns them over and upside down, thereby standing value on its head by substituting a 'fantastic form' for real wood and labor" (217). "Such figures," she adds, "bend toward allegory because they are seemingly animated representations of abstract ideas" (218). This kind of animation is also marked in writers at mid-century with no affinity for Chernyshevsky at all.

In *Reinventing Allegory* (1997), Kelley devotes considerable space to Eliot, not in *Middlemarch*, but in *Daniel Deronda*. Like Jeremy Tambling in *Allegory* (2009), she also focuses on Dickens. In *Bleak House*, Dickens, like Marx, brings the inanimate to startling and grotesque life, as in the houses that "frown" (Dickens 680) at Esther, or, most strikingly, in the case of Tom-All-Alone's. On this street, Dickens writes:

> ruined shelters have bred a crowd of foul existence that crawls in and out of gaps in walls and boards; and coils itself to sleep, in maggot numbers, where the rain drips in; and comes and goes, fetching and carrying fever, and sowing more evil in its every footprint, than Lord Coodle, and Sir Thomas Doodle, and the Duke of Foodle, and all the fine gentlemen in office, down to Zoodle, shall set right in five hundred years – though born expressly to do it. (217)

Like Eliot, only with greater frequency and far more explicit meaning, Dickens also relies on the allegorical force of names, not just Lords Coodle and Doodle, but Blaze and Sparkle the jewelers, Sheen and Gloss the mercers, and even Honoria, Lady Dedlock, the lady who lost her honour long ago. Kelley argues that Eliot's leanings towards allegory in *Daniel Deronda* reflect a broader liberal disappointment after 1848 that left "realism's presentation of particulars and individuals ... less convinced about its representational status (Kelley 218). For Tambling, allegory in Dickens, as in Marx and Baudelaire, responds instead to the rise of the modern city. In his words, "[t]he nature of the city is not that it cannot be read, but rather that its signs are too many and plural, so that there is no adequate way of reducing it to a single meaning" (Tambling 100–1). In the case of *Bleak House*, however, allegory also points directly to the Peircean practice of abduction.

Despite its contribution to an emerging genre of the detective novel, few readers discern in *Bleak House* a particularly scientific bent, and Dickens himself is not very persuasive when he asserts the scientific validity "of what is called Spontaneous Combustion" (Dickens xxxiv) in the face of the many protests he has received from the scientific community, including from his "good friend MR. LEWES." Dickens himself nonetheless makes a clear connection between allegory and the scientific method that unlocks the mystery of the two deaths at the heart of his novel. Inspector Bucket, like Dr. Woodhouse, practises a Sherlock Holmes–ian style of investigation, if admittedly with a supernatural edge; while Snagsby worries all along that Bucket is endowed with "an unlimited number of eyes" (311), at key moments in the novel it is otherwise difficult to understand the powers at his command. Most notably, as Bucket waits for Esther before setting off in search of the missing Lady Dedlock:

> he mounts a high tower in his mind and looks out far and wide. Many solitary figures he perceives creeping through the streets; many solitary figures out on heaths, and roads, and lying under haystacks. But the figure that he seeks is not among them. Other solitaries he perceives, in nooks of bridges, looking over; and in shadowed places down by the river's level; and a dark, dark, shapeless object drifting with the tide, more solitary than all, clings with a drowning hold on his attention. (753)

It would indeed seem, as the narrator also warns, that "[t]ime and place cannot bind Mr. Bucket." Bucket's supernatural trappings are nonetheless made one with the real-world and indexical workings of abduction when we learn that he possesses a "familiar demon" in the form of a "fat forefinger." As the narrator explains, "[t]he Augurs of the Detective Temple invariably predict that when Mr. Bucket and that finger are in much conference, a terrible avenger will be heard of before long" (698). What Mr. Bucket's finger does, of course, is point, as does Allegory on the ceiling of Mr. Tulkinghorn's office.

Tulkinghorn is in many respects the villain of the piece, as the Dedlock family lawyer and keeper of the family secrets who drives Lady

Dedlock to her death; his murder is also the last of the mysteries that Inspector Bucket is called upon to solve. In a marvellous play of the indexical, Dickens accordingly equips Tulkinghorn's chambers with a highly unusual decoration: a painted representation of Allegory. As the narrator explains, Mr. Tulkinghorn lives in a large house, formerly a house of state:

> It is let off in sets of chambers now; and in those shrunken fragments of its greatness, lawyers lie like maggots in nut. But its roomy staircases, passages, and antechambers, still remain; and even its painted ceilings, where Allegory, in Roman helmet and celestial linen, sprawls among balustrades and pillars, flowers, clouds, and big-legged boys, and makes the head ache – as would seem to be Allegory's object always, more or less. (Dickens 130–1)

For all its pseudo-classical trappings, this image of signification has little to do with allegory as Fletcher et al. describe it, as an arbitrary sign firmly and clearly under the control of the artist. While Tambling argues that "Allegory's finger points down to nothingness, makes a gesture of signification but is signifying nothing" (Tambling 105), it is perhaps more accurate to say that the sign on Tulkinghorn's ceiling points to the act of detection that is the lawyer's as it is Bucket's main function. Tulkinghorn's relentless investigation of Lady Dedlock motivates much of the plot, so much so that the repeated mention of his unusual ceiling decoration, like the canary perched on the forehead of the otherwise terrifying Mr. Boythorn, serves as an emblem of his inner being. While the Inspector in any case is introduced into the novel under Tulkinghorn's auspices, his appearance is also framed by the same act of pointing. As the chapter that presents Bucket begins: "Allegory looks pretty cool in Lincoln's Inn Fields, though the evening is hot; for, both Mr. Tulkinghorn's windows are wide open, and the room is lofty, gusty, and gloomy" (Dickens 301). The emphasis on pointing already suggests the indexical qualities that inhere in detection. That association is made starkly apparent, however, when the painted Roman's pointing finger indicates an actual crime.

When novelistic reality clicks into place, Allegory finds himself pointing as a curiously Dickensian version of Peirce's "piece of mould with a bullet-hole in it as a sign of a shot." As a mysterious woman approaches the street sweep Jo apparently right outside Tulkinghorn's window, Allegory gives notice. "From the ceiling," the narrator writes, "foreshortened allegory, in the person of one impossible Roman upside down, points with the arm of Samson (out of joint, and an odd one) obtrusively toward the window." "Why should Mr. Tulkinghorn, for such no-reason, look out of the window?" the narrator asks. "Is the hand not always pointing there? So he does not look out of window" (Dickens 220). For all his sharp powers of observation, Tulkinghorn again fails to heed that pointing finger when it is not Lady Dedlock in disguise, but Lady Dedlock's former maid who "hovers" just outside. While Snagsby has already given warning, Allegory also tries, not that Tulkinghorn "honor[s] him with much attention." As the narrator explains, "[i]t is too dark to see much of allegory over-head there, but that importunate Roman, who is for ever toppling out of the clouds and pointing, is at his old work pretty distinctly" (575). Tulkinghorn pays dearly for this lapse: when Allegory next points, it is to Tulkinghorn's own dead body.

As Tulkinghorn returns from the Dedlocks for the very last time, "nothing meets him, murmuring, 'Don't go home!'" (Dickens 648). Even as he enters his chambers and looks up to "see the Roman point from the ceiling, there is no new significance in the Roman's hand to-night ... to give him the late warning, 'Don't come here!'" (649). Once a shot rings out, however, that meaning is made. "For many years," the narrator writes:

> the persistent Roman has been pointing, with no particular meaning, from that ceiling. It is not likely that he has any new meaning in him to-night. Once pointing, always pointing – like any Roman, or even Briton, with a single idea. There he is, no doubt, in his impossible attitude, pointing, una-vailingly, all night long. Moonlight, darkness, dawn, sunrise, day. There he is still, eagerly pointing, and no one minds him.
>
> But, a little after the coming of the day, come people to clean the rooms. And either the Roman has some new meaning in him, not expressed before, or the foremost of them goes wild; for, looking up at his out-stretched hand, and looking down at what is below it, that person shrieks and flies. (650–1)

When a body lies beneath that ceiling, Allegory gains a specific meaning that is neither arbitrary nor pre-assigned, but instead enjoys an "existential" relationship to what it represents as "a real thing or fact which is a sign by virtue of being connected with it as a matter of fact." Now the painted Roman:

> is pointing at a table, with a bottle (nearly full of wine) and a glass upon it, and two candles that were blown out suddenly, soon after being lighted. He is pointing at an empty chair, and at a stain upon the ground before

it that might be almost covered with a hand.... So, it shall happen surely, through many years to come, that ghostly stories shall be told of the stain upon the floor, so easy to be covered, so hard to be got out; and that the Roman, pointing from the ceiling, shall point, so long as dust and damp and spiders spare him, with far greater significance than he ever had in Mr. Tulkinghorn's time, and with a deadly meaning. For, Mr. Tulkinghorn's time is over for evermore; and the Roman pointed at the murderous hand uplifted against his life, and pointed helplessly at him, from night to morning, lying face downward on the floor, shot through the heart. (651)

In its more conventional rendering, allegory practises a particularly fixed and one-way meaning-making: as J. Hillis Miller argues, the "true" meaning of allegory is one, and it most often lies outside the text. As a pointing finger that indicates meaning that both is and yet is entirely circumstantial, however, allegory in Dickens operates in a very different fashion. Neither side of Dickens's figure, pointing finger or dead body, is primary. Instead, each side completes the other in an indexicality that is no less true for its radical contingency and even uncertainty. While allegory as Allegory, the painted Roman on Tulkinghorn's ceiling, is a stunning image in its own right, it also opens up the possibilities for the device more generally, including as practised by yet another writer with a marked affinity for Dostoevsky: Nathaniel Hawthorne. Even as Hawthorne expresses the same "mutuality and reflexivity" as does Dickens, his rendering also returns us to the Chernyshevskyan model, as allegory in *The Blithedale Romance* (1852) is not an attribute or image of detection, but instead a direct if intentionally ambivalent response to the siren song of social utopia.

Allegory and the Instability of Meaning

Writing in the belated literary culture of North America, Hawthorne, allegories and all, is usually read with reference to Romanticism, and Géza Horváth makes the case for his influence on Dostoevsky in exactly those terms. Horváth draws on Hawthorne's significant presence in the Russian "thick" journals throughout the 1850s to argue that Dostoevsky's overt allegorization in *Crime and Punishment*, including his use of particular word and image clusters (the New Jerusalem, for example, or the motif of treasure), borrows directly from Hawthorne's experimentation with the genre of romance in *The Scarlet Letter* (1850). For all his romantic leanings, however, Hawthorne wasn't immune to literary currents at mid-century, and, with Dickens in mind, allegory in

The Blithedale Romance offers itself up to a different reading. As Horváth notes, while *The Blithedale Romance* wasn't published in Russian until 1905, the files associated with the Dostoevsky brothers' 1864–5 journal *Epoch* contain a draft of a Russian translation (Horváth 33). Suggestive as that finding may be, the importance of the novel for a consideration of Dostoevsky's own use of allegory lies less in any claim of direct influence than in the simple fact of shared content. Like all of Dostoevsky's writing post-Siberia, *The Blithedale Romance* is Hawthorne's often mocking account of a distinctly Chernyshevskyan enterprise, Brook Farm, the real-world experiment in communal living that Hawthorne himself briefly joined in 1841.

In *The Blithedale Romance* as in *The Woman in White*, the materialist claims of "vulgar" science are cast in terms of a mesmerism that Hawthorne flatly rejects, as Hawthorne gives no very good account of the socialist utopian project as a whole. His narrator, Miles Coverdale, has little positive to say about his own motivations for joining the intentional community at Blithedale, nor about that of his friends: the vibrantly beautiful Zenobia, modelled in many respects, as Hawthorne's contemporaries noted, on the real Margaret Fuller; the shadowy Priscilla, who turns out to be Zenobia's real half-sister; and the "philanthropist" Hollingsworth, who seeks to destroy Blithedale for his own ends. Together with the mysterious Westervelt, the four turn the aspiration for social utopia into a tawdry story of power, money, and dysfunctional families that ends with Zenobia's sensational and overwrought suicide, Priscilla's continued subservience now to another man, Hollingsworth's collapse, and Coverdale's retreat to Boston. Along the way, Coverdale makes a point of ridiculing whatever Fourierist impulse underlay the Blithedale experiment, starting with his perusal of what he calls "a series of horribly tedious volumes" immediately upon his arrival at the farm. As Coverdale mockingly summarizes their contents for Hollingsworth:

> "When, as a consequence of human improvement," said I, "the globe shall arrive at its final perfection, the great ocean is to be converted into a particular kind of lemonade, such as was fashionable at Paris in Fourier's time. He calls it *limonade à cèdre*. It is positively a fact! Just imagine the city-docks filled, every day, with a flood-tide of this delectable beverage!" (Hawthorne, *Blithedale* 49)

Despite his flippancy throughout, however, and despite even the catastrophe that ensues, Coverdale is not actually opposed to the project of social utopia, only to certain flawed versions.

Hollingsworth's response to Coverdale already suggests that distinction:

> "Let me hear no more of it!" cried he, in utter disgust. "I never will forgive this fellow! He has committed the Unpardonable Sin! For what more monstrous iniquity could the Devil himself contrive, than to choose the selfish principle – the principle of all human wrong, the very blackness of man's heart, the portion of ourselves which we shudder at, and which it is the whole aim of spiritual discipline to eradicate – to choose it as the master-workman of his system?" (Hawthorne, *Blithedale* 50)

The irony is that Hollingsworth himself operates on the "selfish principle," as Zenobia, again ironically given her own self-absorption, but nonetheless correctly proclaims at the novel's denouement: "It is all self!... Nothing else; nothing but self, self, self!" (201). The original Blithedale, however, was a different proposition, as Coverdale remembers with some wistfulness in the novel's final pages. "Often ... in these years that are darkening around me," he writes, "I remember our beautiful scheme of a noble and unselfish life, and how fair, in that first summer, appeared the prospect that it might endure for generations, and be perfected, as the ages rolled away, into the system of a people, and a world" (225–6). "[M]ore and more," he adds, I feel that we had struck upon what ought to be a truth. Posterity may dig it up, and profit by it" (226). While scholars struggle to account for Hawthorne's six-month stay at the real Brook Farm, the admittedly unreliable Coverdale may offer some insight here. Fourier may not be it, and Hollingsworth with his plan to turn Blithedale literally into a prison even less so. A real Arcadia where Zenobia and Priscilla are truly sisters and matter and spirit equal players nonetheless remains an ideal, a complicated materiality that Hawthorne expresses above all in an allegory marked by its ambivalence.[7]

Hawthorne himself acknowledged an "inveterate love of allegory" (Hawthorne, *Complete* 1043), perhaps most evident in Hester Prynne's gaudy scarlet "A." In *The Blithedale Romance*, we might note the over-determined symbolism of Priscilla's veil and Zenobia's flower, for example, as in the often openly artificial names of Hawthorne's characters, including his narrator. In contrast to Chernyshevsky's all-knowing narrator, Hawthorne's unreliable Coverdale covers more than he reveals, and he is often simply ill-informed, with a tendency to not quite catch important conversations or to arrive just after a climactic scene; he is also not in very good control of the signs that, like his name, point in too many directions at once.[8] The result of all this insistent pointing, exactly unlike Chernyshevsky's, is an instability of meaning anchored in Hawthorne's refusal to prioritize one level of meaning over another.

Early in his narrative, for example, Coverdale returns over several pages to the fire that on various levels welcomed him to Blithedale and warms him still. "There can hardly remain for me," he begins,

> (who am really getting to be a frosty bachelor, with another white hair, every week or so, in my moustache,) there can hardly flicker up again so cheery a blaze upon the heart, as that which I remember, the next day at Blithedale.… Vividly does that fireside re-create itself, as I rake away the ashes from the embers in my memory, and blow them up with a sigh, for lack of more inspiring breath. (Hawthorne, *Blithedale* 9)

At this point Coverdale interrupts his story to describe his actual journey to Blithedale, only to then return to the fire, which is now not just real and remembered but also literary. "And, now," he says, "we were seated by the brisk fireside of the old farm-house; the same fire that glimmers so faintly among my reminiscences, at the beginning of this chapter" (12). As Coverdale again abruptly departs from the scene at hand to sketch in the larger ideals of the group, the fire serves him yet again:

> Therefore, if we built splendid castles (phalansteries, perhaps, they might be more fitly called,) and pictured beautiful scenes, among the fervid coals of the hearth around which we were clustering – and if all went to rack with the crumbling embers, and have never since arisen out of the ashes – let us take to ourselves no shame. (19)

The real fire that is also, as Coverdale later says, the illumination of Blithedale's morals as it is of his own memories, is finally the warmth of one human being for another, as Hollingsworth arrives with a very chilled Priscilla and warns, "The very heart will be frozen in her bosom, unless you women can warm it, among you, with the warmth that ought to be in your own!" (27). One kind of fire, or frost, is not more real than any other, as Hawthorne's prolonged oscillation between the two halves of his figure upends the hierarchy of meaning that marks a more standard sort of allegory as it does garden-variety social utopianism. Dostoevskyan allegory produces the same effect for the same reasons, above all in *Demons*.

If all Dostoevsky's work post-Siberia takes on Chernyshevsky's social-utopian project, *Demons* with its tale of the cataclysmic chain of events set in motion by two generations of would-be revolutionaries is his most direct response. While Dostoevsky, as always, presents death as the inevitable outcome of nihilist thought, in *Demons* he is also explicit in pinning the blame on what his hero Shatov calls "half-science." "Half-science," Shatov tells Stavrogin, "is a despot such as has never been seen before. A despot with its own priests and slaves, a despot before whom everything has bowed down with a love and superstition unthinkable till now, before whom even science itself trembles and whom it shamefully caters to" (*Demons* 251; *PSS* 10: 199). Dostoevsky's attack is at its most pointed when conveyed in the mockery of small details, as in the "progressive" picnic that devolves into a drunken cancan, or the attempt at open marriage that sets mind and body of the aptly named Virginsky at odds. It is also strikingly apparent in an allegorical practice that, like Hawthorne's, first parodies and then entirely re-casts allegory of the "vulgar," Chernyshevskyan sort.

When our chatty narrator in *Demons* insists that his friend Stepan Trofimovich actually pounded the walls in frustration, he explains, "This occurred without a trace of allegory, so that once he even broke some plaster from the wall," while one of the guests at the ill-fated fête takes a little too much care to explain what would seem a very ordinary figure of speech. "I am speaking al-le-gor-i-cally," he clarifies, "but I went to the buffet and am glad to have come back in one piece." "These are all nonsensical allegories," an angry Varvara Petrovna tells Lebiadkin, "These are allegories, and, besides, you choose to speak too floridly" (*Demons* 14, 506, 176; *PSS* 7: 15, 528, 187). Along with an insistence on a particularly flat kind of matter as the only measure of reality, their usage reflects a gnawing concern that the "real" significance of words lies elsewhere, a concern entirely appropriate to a novel where double meanings run rampant. Unfortunately, the characters' recourse to "allegory" only makes a bad situation worse.

While Shatov as a recovering revolutionary is largely innocent of allegorical intention, he is nonetheless not entirely immune to the practice of figuration, for example, when he refers to his serfdom as both literal and metaphorical. "Once I was simply born of a lackey, but now I've become a lackey myself, just like you," Shatov says. "Our Russian liberal is first of all a lackey and is only looking for someone's boots to polish." The narrator's immediate turn to "allegory" doesn't clarify Shatov's meaning, however, but only indicates his own or perhaps also our inability to see where that figure might lie: "What boots?" Anton Lavrentievich asks. "What kind of allegory is that?" (*Demons* 138; *PSS* 7: 538). Fortunately, even as Dostoevsky's heroes repeatedly generate not just more confusion, but finally even death in their attempt to reduce the multiplicity of "living life" to a single level of reality, another option remains available to his readers. Rather than remain mired in the non-coincidence of sign and signified that so afflicts Dostoevsky's heroes, like Dickens and like Hawthorne, we can accept instability instead.

Russell Valentino notes what he calls the novel's "ambivalent orientation towards its own allegorical status," an ambivalence apparent even at the fête, where words, as it turns out, do have tangible effects. As the now truly mad von Lembke rightly says, "Governesses have been used to set houses on fire…. The fire is in people's minds, not on the rooftops" (Valentino 117; *Demons* 516; *PSS* 7: 538).[9] When we cast mind as part of a material world that is itself multiple and even shifting, however, it is not so much ambivalence as it is two different kinds of allegories in operation at once. Where his characters repeatedly attempt an allegory of the familiar, "vulgar" sort, Dostoevsky himself uses allegory to point to multiple "real" realities that are simultaneously also symbolic. This play of clearly defined and yet multiple and even contradictory meanings operates most aggressively at the novel's very end, when Stepan Trofimovich attempts to apply the parable of the Gadarene swine to his own Russian reality.

As his Bible-selling companion at his request reads from the Gospel according to Luke, Stepan Trofimovich is struck by what he calls *"une comparaison."* "It is us, us and them, and Petrusha … *et les autres avec lui*, and I, perhaps, first, at the head, and we will rush, insane and raging, from the cliff down into the sea, and all be drowned," he says. "But the sick man will be healed and 'sit at the feet of Jesus'" (*Demons* 655; *PSS* 7: 681). The *"comparaison"* proves a happy one, not least because it turns out to have shaped our reading all along. As the narrator now takes a moment to remind us, the passage from Luke that Stepan Trofimovich finds so meaningful is the very one that the narrator himself placed at the beginning of the novel. While Stepan Trofimovich and the narrator together seem to have unlocked the allegory that is the novel as a whole, the very circularity of our own evidently over-determined reading renders that meaning a little suspect. Up to the very end, with his Gallicisms and his inveterate "quotation," Stepan Trofimovich also remains the deeply untrustworthy wielder of words that he has been all along. To the dismay of the "perspicacious" reader of the Chernyshevskyan type, as the allegorical interpretation of *Demons* as those demons, the ones from Luke, is both offered and withheld, our way forward is neither easy nor open to debate. As the equally unreliable "Petrusha" says twice, *"Avis au lecteur"* (359; *PSS* 7: 377): our only choice is to read the novel both ways at once.

The mistake of "vulgar" materialism is the reification of what, on closer look, reveals itself to be a largely ideological "abstraction from objects of the senses." As Chernyshevsky's allegory exactly reflects, while material monism lays claim to a single objective reality, its very refusal to acknowledge the mutual implication of mind and matter makes duality inevitable. Allegory in Dostoevsky, on the other hand, posits multiple meanings in operation at the same time and with the same degree of "objective" reality. Exactly like Lewes's nerves and neuroses, and as challenging as it may be for Dostoevsky's readers, even two diametrically opposed readings reflect "a common identity of pressure": in Dostoevsky's allegory, as in his metaphor, inside and outside are one, as "the convex and concave surfaces of the same sphere, distinguishable yet identical." The oscillation between different expressions of what is nonetheless the same reality that in Dostoevsky approaches a kind of whiplash suggests a particularly demanding form of romantic irony, which is to say, Dostoevsky's project, like Hawthorne's, shares a great deal with Goethe's and Coleridge's. The indexicality that marks even his allegory, however, also makes the case for what Latour calls "a more 'realistic realism'" (Latour 15).

In an 1868 letter to his friend Apollon Maikov, Dostoevsky himself acknowledged that he held "completely different notions of reality and realism than our realists and critics" (*PSS* 28/2: 239). As Liza Knapp argues, Dostoevsky claimed more than once that the distinctive feature of his realism was his commitment to presenting reality "as *he* experienced it" (Knapp, "Realism" 235); for Molly Brunson, Dostoevsky's "realism in a higher sense" offers a "transcendent alternative to a more grounded, objective recording of phenomenal reality, one capable of accessing truths far higher, or deeper, than those of the material world" (Brunson 163). We needn't cut off from the material world, however, to find truths "far higher, or deeper" than Wellek's "orderly world of nineteenth-century science" would allow. For an entirely mainstream but apparently less orderly nineteenth-century science, the material world manifests itself in multiple ways, including in "fantastic" perceptions and in ways altogether beyond the reach of our particular perceptual apparatus. In Menand's terms, this world is one always susceptible to "'a certain swerving,'" Even in our own twenty-first century, the lessons of this other sort of science don't come easily, but they restore the figurative possibilities of language as they open us to a "living life" that always operates on multiple levels at once. As the association of allegory with and against social utopianism makes especially clear, they finally also open us to a particular kind of politics. As we will see in the next and last chapter, even in the case of the famously and sometimes stridently conservative Dostoevsky, this politics is less a matter of "left" or "right," than it is of a mutual and reflexive way of being in the world that operates at the same time in human terms alone.

5 Social Physics

Midway through *Middlemarch*, one of the locals offers an entirely Dostoevskyan description of Will Ladislaw. Will has now embarked on the newspaper business and also taken up the cause of reform. On being told that his name is Ladislaw and that he is "said to be of foreign extraction," Mr. Hawley says: "I know the sort … some emissary. He'll begin with flourishing about the Rights of Man and end with murdering a wench" (Eliot, *Middlemarch* 358). Even in the more genteel surroundings of Middlemarch, it is the plot of *Crime and Punishment* in brief, not that Mr. Hawley, nor even his far more well-read author, knows the earlier work. By 1871 and the publication of *Middlemarch*, Will is instead already a type, and undergirding Mr. Hawley's apparently extreme response is a conflation of science and politics that the Russian tradition makes blindingly obvious. In the late twentieth and early twenty-first centuries, the "[i]nhuman, reductionist, causal, law-like, certain, objective, cold, unanimous, absolute" science that Latour calls "Science with a capital S" (Latour 229) largely aligns with corporate interests and the political right. In the mid-nineteenth century, it was most often associated with the radical left instead.

Vera Pavlovna's dream in the fourth chapter of *What Is to Be Done?* borrows from Fourier to imagine not just human interactions but the entire natural world transformed by science and technology, including electric light, aluminum furniture, and new irrigation techniques. As desert is changed into fertile fields and machines do all the work, men and women in conditions of abundance and perfect equality busy themselves in lecture halls and libraries and in romantic relationships unencumbered by social convention. Zola's variant is the "scientific age" that "Claude Bernard saw in his dreams":

When that time comes the doctor will be the master of maladies; he will cure without fail; his influence upon the human body will conduce to the welfare and strength of the species. We shall enter upon a century in which man, grown more powerful, will make use of nature and will utilize its laws to produce upon earth the greatest possible amount of justice and freedom. There is no nobler, higher, nor grander end.

"Here is our role as intelligent beings," Zola concludes: "to penetrate to the wherefore of things, to become superior to these things, and to reduce them to a condition of subservient machinery" (Zola 25). In the Russian context, this particular wave of "extraordinary" Science culminated in the Soviet experiment, complete with the draining of the Aral Sea, plans to colonize other planets, and Lenin's famous slogan: Communism = Soviet Power + the Electrification of the Entire Country. In direct response to what we now recognize as an emerging Communist ideal, the nineteenth-century proponents of a more "mutual and reflexive" science tended to political conservatism, sometimes, as in the case of Dostoevsky, apparently very much so. Even in the extreme case of Dostoevsky, however, what is most striking is the extent to which their argument draws from both sides.

Not everyone emerges from this ambivalence with their political integrity intact. Like science in a world always susceptible to "a certain swerving," a politics of multiplicity runs the risk of "the law of higgledy-piggledy," and certainly Hawthorne often stands accused of exactly that. Scholars have long struggled to reconcile Hawthorne's investment in Brook Farm with his open mockery of the Fourierist impulse in *The Blithedale Romance*, not to mention his widely published views on the inefficacy of social reform. As Richard Brodhead argues, neither stance suits Hawthorne the "public functionary," the one who "as customs officer, United States consul, and consultant on Democratic Party patronage arrangements, put himself at the service of the system of official power in its most institutional forms" (Brodhead 95). Brodhead finds order in this confusion only by reverting to the literary terms of irony. In both *The Scarlet Letter* and *The Blithedale Romance*, Brodhead writes, Hawthorne presents two opposing ideas "in full awareness of their contradiction": "[o]n the one hand, that the political is a sphere of delusion, an arena in which men and women act on other motives than they admit to for other ends than they foresee; and on the other, that to fail to be politically engaged is to sacrifice a dimension, to rob the self and its world of the fullest form of their reality" (102). As in science, so in politics, however, and "higgledy-piggledy" is not the only outcome of a commitment to multiplicity and indeterminacy. Writing just a little later and with more of what we might call a realist bent, both Dostoevsky and Eliot combine Hawthorne's romantic irony with a Feuerbachian "sensuousness" that lends the "fullest form of reality" a more solid material presence. They are also more generous in their estimation of human motives.

Just as Dostoevsky's early dip in the waters of radical thought is evident in his arrest in 1848 together with the other members of the Petrashevsky "circle," so Marian Evans's decision to make her way as a writer in London first on her own and then together with an already-married man, George Henry Lewes, makes her an English incarnation of the "nihilist girl," a real-life version of Vera Pavlovna. Eliot was also an early British advocate of Comte, as well as the English translator of David Strauss's *The Life of Jesus, Critically Examined* (*Das Leben Jesu, kritisch bearbeitet*, 1846) and Feuerbach's *The Essence of Christianity*. Although both Dostoevsky and Eliot then ended their careers as pillars of their respective literary establishments, it was without entirely renouncing their more radical beginnings. Both instead, in Eliot's words, warn of the dangers of a "binding theory" (Eliot, *Middlemarch* 86) that fails to take into account the existence of real others, while still paying tribute to the ardour of young people with utopian plans for re-making the world. If their evolving life experiences surely inclined them in that direction, the kinder eye that Dostoevsky and Eliot cast on the political "sphere of delusion" is finally more than a matter of biography alone. Both were also devoted readers of Schiller.

Poet, playwright, and philosopher Friedrich Schiller famously inspired several generations of Russian revolutionaries and reformers, including Dostoevsky himself in his early, more radical phase. As Edward K. Kostka writes, Schiller was "the principal source of his youthful exaltation," "the poet whom Dostoyevsky idolizes during his student years" (Kostka 215). In the more politically stable conditions of Great Britain, Eliot admired him with more restraint but in exactly the same terms; as Donald Stone writes, "Schiller fired her with a devotion to lofty feeling" (Stone 188). While many of Schiller's nineteenth-century admirers never moved past "youthful exaltation," the adventures of their ardent young heroes suggest that Dostoevsky and Eliot read Schiller a little more attentively. In Karl Moor in *The Robbers* and the Marquis de Posa in *Don Carlos*, idealistic fervour runs the risk of a one-sidedness or even fanaticism that Schiller himself called "Schwarmerei," not that he rejected ardour altogether. The way forward is instead a recalibration that Eliot's narrator in *Middlemarch* suggests when she sums up the fundamental failing of the intensely evangelical Bulstrode: "There is no general doctrine which is not capable of eating out our morality if unchecked by the deep-seated habit of direct fellow-feeling with individual fellow-men" (Eliot, *Middlemarch* 619). At the risk of a large dose of old-fashioned sentimentality, in Dostoevsky and Eliot as in Schiller, we resolve the contradictions of ardour when we turn our "lofty feelings" towards even just one real person in the world.

Writing *The Gift* on the other side of the Russian Revolution and in an era of high modernism, Nabokov no longer puts much stock in ardour. In Nabokov's twentieth-century Berlin, parallel tram lines meet and the mutual implication of mind and material world is evident even in so small a detail as a mirror unloaded from a moving van "across which, as across a cinema screen, passed a flawlessly clear reflection of boughs sliding and swaying not arboreally, but with a human vacillation, produced by the nature of those who were carrying this sky, these boughs, this gliding façade" (Nabokov 6). Above all, and in direct opposition to Chernyshevsky's "vulgar," purblind, and proto-Marxian materialism, *The Gift* shows us a world where both natural and man-made objects are endowed with their own agency: "the peeled walls of old houses toasting their tattooed backs in the morning sunshine" (328), a greengrocery that crosses the street "with a glance over its shoulder" (5), animals and plants alike marked by "the incredible artistic wit of mimetic disguise" (110). Along with the nineteenth-century commitment to plot as Brooks describes it, what has gone by the wayside is "the deep-seated habit of direct fellow-feeling with individual fellowmen." Nabokov's lack of sympathy for Chernyshevsky and his nihilist friends is total, and, if love prevails, still Fyodor's relationship with Zina is a little too straightforwardly allegorical, a little too much about Nabokov's own art-in-the-making, to offer much in the way of a political stance. To the extent that the heroes of the nineteenth-century novel survive their experience of ardour, however, it is because they find one another, translating the dialogic interrelationship of mind and material world into human terms alone.

Nihilist Science and Left-Wing Politics

The nineteenth-century purveyors of "Science with a capital S" didn't always advocate radical left-wing politics. To the extent that he engaged in politics, Claude Bernard, for example, allied himself instead with the aims of the French imperial state. As Jerome Tarshis writes, in 1869 Napoleon III appointed him to a seat in the senate where he "seems to have been ideally suited to be a rubber-stamp legislator." Far from disrupting the status quo, Bernard apparently only once undertook to propose new legislation, and then only to cede still more power to the emperor. As Tarshis explains, "[t]ogether several other senators, he sponsored a constitutional amendment giving the emperor the power to appoint senators without consulting anyone at all" (Tarshis 136). Cell biologist Rudolf Virchow, on the other hand, was an active legislator along what we might call liberal lines. After taking to the barricades in 1848, Virchow was elected to the Berlin city council in 1859, in 1860 he joined the Council of Scientific Advisors to the Prussian government, in 1861 he was elected to the Prussian Diet, and from 1880 to 1893 he served in the Reichstag. Throughout, as Laura Otis writes, Virchow's highly successful promotion of public health as a social issue, as well as his resistance to Bismarck's anti-constitutional consolidation of power, re-cast his scientific work on cells as "the bourgeois ideal of the free, responsible, and 'self-contained' individual" (Otis 9). The less the scientist-popularizers of the nineteenth century engaged in actual scientific work, however, the more starkly their politics leaned left.

Of Engels's "vulgar itinerant preacher" materialists, only Moleschott enjoyed a position in the scientific establishment, and that at the direct cost of his early left-wing politics. When the charge of "infecting the minds of the youth" forced Moleschott's resignation from the University of Heidelberg in 1854, he shifted gears and won a highly coveted position in anatomy and physiology at the University of Zurich, due largely, as Frederick Gregory explains, to a positive recommendation from Rudolf Virchow (Gregory 97). Once in Zurich, Moleschott befriended Georg Herwegh and Franz Liszt and entertained various luminaries, including George Henry Lewes and George Eliot. As his scientific interests then waned, Moleschott moved to Italy, first to Turin and then, on his own appointment to the Senate, to Rome, where he successfully rested on his laurels. As Gregory writes, despite his lack of any recent scientific work, "on the celebration of his seventieth birthday in 1892 he received congratulations from great names of nineteenth-century science such as Du Bois-Reymond, Helmholtz, and others" (Gregory 99). Vogt and Büchner as the other two members of the "vulgar" triumvirate, in contrast, traded science for left-wing politics almost immediately.

Vogt first studied chemistry with Liebig himself until his participation in student protests forced him to flee to Bern in 1835. After shifting to zoology, Vogt then spent five years working as an assistant to Swiss zoologist and geologist Louis Agassiz (1807–73) before setting off for Paris in 1844, where he shared a room first with Quetelet and then with none other than Russian anarchist Mikhail Bakunin (1814–76). Although the latter two remained friends until a final political break in 1868, their experiment in shared living lasted a mere two weeks before Bakunin, as Gregory notes, "out of money and credit, spent funds that came in the mail for Vogt, payment for his work as a correspondent for a German periodical. This, plus Bakunin's obsession with Hegel, soon drove Vogt out" (Gregory 228, fn 34). In 1846, Vogt wrote his most famous line:

> Every natural scientist who thinks with any degree of consistency at all will, I think, come to the view that all those capacities that we understand by the phrase psychic activities (Seelenthätigkeiten) are but functions of the brain substance; or, to express myself a bit crudely here, that thoughts stand in the same relation to the brain as gall does to the liver or urine to the kidneys. (64)

In the same year, he returned to Germany to take up a position in zoology, just in time to play a leading role in the German revolutionary movement. Büchner enjoyed a similarly peripatetic life in science for similar reasons and even in direct association with Vogt, for example, in his active support of Vogt's election to the Frankfurt parliament in 1848. When Büchner's extreme views lost him his position as a lecturer in medicine at Tübingen, he was unable to gain another university job, and he instead lived off his writing and a desultory private medical practice in Darmstadt, ending, like Vogt, largely active in politics and filled with bitterness towards a German university system that finally had no place for either (Gregory 70). Comte, as we may recall, lost his chance at a scientific career when he was expelled from the École polytechnique, while the leading Russian nihilists, Chernyshevsky, Dobroliubov, and Pisarev, freely substituted *haute vulgarisation* for any scientific training whatsoever. As striking as these examples are, still the nineteenth-century association between Science and radical politics is at its most pointed when the scientists are not so much under-qualified, as fictional altogether.

Even this claim calls for a degree of qualification. Dickens's Dr. Woodhouse, despite his proto-Holmesian detective work, is a bourgeois promoter of social well-being along the lines of Virchow; while Holmes's own functioning in political as in scientific terms is a matter of some debate, certainly his author, the real Conan Doyle, aimed to serve empire instead.[1] Left-wing politics in the nineteenth century, as now, were also more often a lifestyle choice than actual advocacy of revolution. As Jurij Lotman famously argues with regard to the Decembrists, it is possible to talk about the nihilist "not only as the bearer of a particular political programme, but also as a specific cultural-historical and psychological type" (Lotman 73). Bazarov in *Fathers and Sons*, for example, offers no political plan of action whatsoever as his particular brand of nihilism is defined entirely in terms of what it is not. To quote his friend Arkady once more, "A nihilist is a man who doesn't acknowledge any authorities, who doesn't accept a single principle on faith, no matter how much that principle may be surrounded by respect" (Turgenev 23). The very refusal to acknowledge authority, however, together with Bazarov's own plebeian origins, his overtures to the Kirsanovs' former serfs, and his apparent hostility to Pavel Petrovich's aristocratic ways gesture at a revolutionary movement more fully realized in *What Is to Be Done?* Given the conditions of Russian censorship, especially as Chernyshevsky experienced them while writing and publishing his novel from within the confines of the Peter-Paul Fortress, even *What Is to Be Done?* presents the program of revolution a little obliquely, as a matter again of the heroes' abrupt manners and scientific inclinations, now combined with their unconventional living arrangements and Rakhmetov's apparently clandestine travels across Europe and North America until that time, as the narrator says, when "it would be 'necessary' for him to be in Russia" (Chernyshevsky, *What* 291). As a real revolutionary movement emerges in Russia, Dostoevsky makes the connection between a certain kind of science and revolutionary politics more overtly, albeit still within the confines of what the censor would allow.

This association is especially evident in *Demons* as Dostoevsky's only portrayal of an actual, if utterly inept, revolutionary cell. Dostoevsky's 1871 novel reworks the so-called Nechaev affair, the real murder of a student at the Petrov Agricultural Academy in Moscow in the summer of 1869, and his hero Shatov is explicit in associating the practice of "half-science" with a nascent terrorist movement. More often and across all his novels, however, Dostoevsky presents the political implications of nihilist science as not so much inadequate, as not yet fully realized. In *The Brothers Karamazov*, Rakitin and Kolia Krasotkin amplify Ivan's combined scientific and political tendencies, Rakitin with his chemistry, his talk of "race and selection," and his claims to love "liberty, equality, fraternity" (*Brothers* 82; *PSS* 14: 76), and Kolia with his still more direct, although youthful claims to left-wing politics. "I only respect mathematics and natural science" (551; *PSS* 14: 497), the twelve-year-old Kolia tells Alyosha, followed shortly by, "I'm a socialist, Karamazov, I am an incorrigible socialist" (554; *PSS* 14: 500). In *Crime and Punishment*, a window abruptly opens onto revolutionary politics when a possibility occurs to Raskolnikov's friend Razumikhin. While Dostoevsky's readers may already have recognized the type, Razumikhin draws his own conclusions in one of those brief moments when the narrative perspective shifts to take in the thoughts of a character other than Raskolnikov himself. At a loss to account for his friend's strange words, Razumikhin makes an easy mistake. "He's a political conspirator! For sure! And he's about to take some decisive step – for sure!" (*Crime* 444; *PSS* 6: 340) he thinks to himself. In contrast to Dostoevsky, Eliot didn't have a revolution to anticipate, and *Middlemarch*, while published in 1871, is set in the late 1820s and early 1830s, a little too early for a science, and hence politics, of an explicitly nihilist kind. Even in *Middlemarch*, though, radical politics hover just off-stage.

As Eliot was well aware, the people and events of the 1820s and 1830s when the novel is set resonated very differently in the 1860s and 1870s when the novel was written and published, and while Gillian Beer notes the effect of the second Reform Bill in 1867 on readers' perceptions of the first in 1832, with Dostoevsky in mind we recognize in Lydgate a precursor to the nihilist doctor along the later lines of Claude Bernard or Chernyshevsky's fictional Kirsanov. Lydgate when we meet him is recently returned from France where he engaged in "galvanic experiments" with rabbits and frogs. Now back in England his ultimate goal is to pursue not patients, but research, as he is "ambitious above all to contribute towards enlarging the scientific, rational basis of his profession" (Eliot, *Middlemarch* 147). Eliot makes explicit the political implications that for now are only a possible consequence of Lydgate's sort of science. If Lydgate, we are told, "[i]n warming himself at French social theories … had brought away no smell of scorching," it is because Eliot has made her point and because with Lydgate she is careful to keep within the confines of her historical setting (348). As Hawley's extraordinary statement would indicate, however, with Will, Eliot doesn't always maintain the veneer of historicity.

Despite Will's liberal notions and his partly Polish background, the claim that Will is a murdering spy is not just utterly misplaced but even, as Thomas McLean argues, an obvious anachronism, as is Mrs. Cadwallader's concern that Dorothea's marrying Will would be like marrying "an Italian with white mice" (Eliot, *Middlemarch* 490).[2] As noted above, an Italian with white mice can only be our friend Fosco, and Eliot's strikingly deliberate comparison (it's repeated two pages later), like Hawley's later reference to "Any cursed alien blood, Jew, Corsican, or Gypsy," points instead to an association of revolution, crime, and Napoleon that was much more keenly felt by the date of Eliot's actual writing (719). Just as Fosco combines his career as both secret agent and amateur chemist with a striking resemblance to the former Emperor of France, so Hawley's "Corsican" refers to one Corsican in particular, the one Raskolnikov compares himself to when he offers his most concise explanation of his crime: "I wanted to become a Napoleon" (*Crime* 415; *PSS* 6: 433). Razumikhin's belated notion that Raskolnikov might be engaged in political conspiracy echoes Hawley's suspicions of Will, but only as based in facts more firmly on the ground by the 1860s. Raskolnikov is a Will Ladislaw pushed further east but still set in revolutionary opposition to the same imperial power and made more real by the radicalization not just of Russia but of Eliot's own day. While Will is a much milder sort, in short, still his neighbours' fears aren't entirely misplaced. Indeed, like Dostoevsky and to a lesser extent Hawthorne, Eliot isn't wholly opposed to the desires of Chernyshevsky's "new people," not that she writes in support of the nihilist program, either. The message that *Middlemarch* sends is instead deliberately mixed, although Eliot approaches that ambiguity from the opposite direction. In *Crime and Punishment*, Raskolnikov, despite his unfortunate philosophical stance, despite even the double murder that he commits, remains Dostoevsky's hero. In *Middlemarch*, Will and Dorothea are more evidently intended to elicit readers' sympathies. For all that she endows them with good looks and obvious good intentions, however, Eliot makes a point of marking her heroes with a degree of "higgledy-piggledy," not just in terms of the rumours that swirl around Will, but above all in the Dostoevskyan effects of Dorothea's attempts to do good.

A Politics of Ambivalence

The radical edge to Dorothea's politics has often been lost on her readers. Despite her wealth and gentility, Dorothea promotes distinctly utopian schemes that in her case tend to revolve first around the design of ideal living spaces and then the creation of what she calls a "Pythagorean community." As she explains late in the novel, "I should like to take a great deal of land, and drain it, and make a little colony, where everybody should work, and all the work should be done well. I should know every one of the people and be their friend" (Eliot, *Middlemarch* 550). By the 1870s there were a number of models for Dorothea's project, including Hawthorne's Brook Farm, Robert Owen's New Lanark and New Harmony, and, not that Eliot knew it, Chernyshevsky's vision in Vera Pavlovna's fourth dream; in Eliot's fictional 1830s, Dorothea's inspiration is the French pastor and philanthropist J.F. Oberlin (1740–1826).[3] While Dorothea's "ardour" (50) and "exalted enthusiasm" (28) lift her above the average citizens of Middlemarch in our estimation, there is often something faintly comical in her earnest attempts to do good, especially as seen through the eyes of her younger sister Celia. As the narrator explains Celia's perspective late in the novel, "[a]ll through their girlhood she had felt that she could act on her sister by a word judiciously placed – by opening a little window for the daylight of her own understanding to enter among the strange coloured lamps by which Dodo habitually saw" (820). When Dorothea turns her energies on her first husband, however, ardour takes a darker turn.

In *Middlemarch* we only dance around the possibility of murder, not just in Hawley's insinuations regarding Will, but more importantly in Lydgate's implication in the death of Raffles, and Dorothea herself would seem impossibly far from Raskolnikov's axe-murdering excess. Still, Casaubon's acquaintance with the determinedly well-intentioned Miss Brooke proves equally fatal. Eliot is quite clear that Dorothea thinks far more about her own imagined role than about the real Casaubon and that there is accordingly something quite oppressive about Dorothea's repeated offers to help. As Nina Auerbach writes, "Personal disclosure: I hate people asking dulcetly when my book will be finished, as Dorothea does incessantly" (Auerbach 92). Dorothea, we are told, "had thought that she could have been patient with John Milton, but she had never imagined him behaving in this way" (Eliot, *Middlemarch* 282). In her frustration Dorothea lashes out, half an hour later the already-ill Casaubon suffers his first heart attack, and soon he is dead.

Dorothea's own awareness that she had pushed her husband too far is evident in her anguished cry to Lydgate: "I beseech you to speak quite plainly ... I cannot bear to think that there might be something which I did not know, and which, if I had known it, would have made me act differently" (Eliot, *Middlemarch* 288). As the narrator explains, there was indeed something she could have done differently:

> We are all of us born in moral stupidity, taking the world as an udder to feed our supreme selves: Dorothea had early begun to emerge from that stupidity, and yet it had been easier to her to imagine how she would devote herself to Mr Casaubon, and become wise and strong in his strength and wisdom, than to conceive with that distinctness which is no longer reflection but feeling – an idea wrought back to the directness of sense, like the solidity of objects – that he had an equivalent centre of self, whence the lights and shadows must always fall with a certain difference (211).

In Porfiry Petrovich's words in *Crime and Punishment*, "it's a dangerous thing in young people, this suppressed, proud enthusiasm!" (*Crime* 452; *PSS* 6: 345). Like so many of Dostoevsky's heroes, not just Raskolnikov and Ivan Karamazov, but Shatov and Kirillov in *Demons* and Ippolit in *The Idiot*, Dorothea is a bright young person with utopian plans for re-making the world. We may cherish her good intentions, and many readers do. As both Porfiry Petrovich and Mr. Hawley warn, however, in the conditions of the real world, left-wing "enthusiasm" can prove not just foolish but actually deadly.

Where Eliot is only ever gentle in implicating Dorothea in the more extreme effects of ardour, Dostoevsky, like Hawthorne, is much more overt. Like the real Hawthorne, Coverdale lasts at Blithedale only briefly before returning to a more conventional existence; although he has his wistful moments, Coverdale's tone as the narrator of the events that make up the novel is also flippant throughout. Dostoevsky's ridicule of the radical left is still more scathing, especially when the joke comes explicitly at Chernyshevsky's expense. Dostoevsky regularly re-purposes images and episodes from *What Is to Be Done?* to wickedly comical effect, when the Crystal Palace reappears as a seedy bar, for example, or when the Underground Man obsessively plans to bump shoulders. His minor characters often also serve to disseminate a particularly silly version of nihilist thought. In *Demons*, for example, "Our People" add to the ludicrous impression produced by their pseudo-Chernyshevskyan picnicking when their so-called meeting runs through a laundry list of nihilist talking points, from women's rights, atheism, and possible government informers to Fourier and aluminium columns. In *Crime and Punishment*, Lebeziatnikov achieves the same effect all on his own, as he touches in the space of just a few pages on everything from "Fourier's system and Darwin's theory" (*Crime* 365; *PSS* 6: 278) to the rights of women, the importance of environment, and even the concept of usefulness. "I understand only the one word: *useful!*," he concludes, "Snigger all you like, but it's true!" (371; *PSS* 6: 285). Just as Hawthorne and Dostoevsky are much broader in their mockery of the socialist utopian impulse, so both are also far more pointed and even gruesome in tracing its deadly effects.

The violence that death entails in both *Crime and Punishment* and *The Blithedale Romance* makes their critique of radical politics especially obvious. In *Crime and Punishment*, the narrator dwells on the blood that pours out of the old pawnbroker's head, onto the string around her neck, and into a pool on the floor. As her body falls, the narrator writes, "[h]er eyes bulged as if they were about to pop out, and her forehead and her whole face were contracted and distorted in convulsion" (*Crime* 77; *PSS* 6: 63). The ghastly detail of the first murder then repeats in the second, as Raskolnikov turns to her sister and his blow lands "directly on the skull, with the sharp edge, and immediately split the whole upper part of the forehead, almost to the crown" (79; *PSS* 6: 65). In *The Blithedale Romance*, Zenobia's body serves as the object of Coverdale's mocking fascination starting with their very first encounter, when Zenobia's reference to "the garb of Eden" together, Coverdale says, "with something in her manner," brings to his imagination "a picture of that fine, perfectly developed figure, in Eve's earliest garment";

as he adds, "I almost fancied myself actually beholding it" (Hawthorne, *Blithedale* 17). Coverdale's mockery returns to haunt us at the novel's end, however, when Zenobia's once-beautiful body is defiled first by her drowning – "Of all modes of death," Coverdale says, "methinks it is the ugliest" (216) – and then by the men's attempts to find it. Silas Foster's memories of stealing out of bed "to go bobbing for horn-pouts and eels" are metaphorically a little crude. "Then, I was a boy," he recalls, "bobbing for fish; and now I am getting to be an old fellow, and here I be, groping for a dead body!" (214). What is worse is the effect of that "groping": when the hook on the pole that they use to dredge the river catches Zenobia's breast, it leaves a wound near her already-dead heart. *The Blithedale Romance* concludes on that dark note, as Coverdale's criticism of social utopianism finally well outweighs his passing acknowledgment of the idealism that drives young people to try to make the world better. In the very over-determination that marks his response to nihilism, as it does his symbolic practice, however, Dostoevsky tips the balance just the other way.

Where Eliot's criticism of her heroes' nihilist tendencies is so attenuated as to pass most readers by, Dostoevsky's insistent and even over-the-top recourse to the most extreme of plot elements, including murder, suicide, and fire, sets his attack on left-wing radical politics front and centre. What accordingly sometimes escapes his readers' attention, in turn, is the extent to which Dostoevsky at the very same time supports certain aspects of nihilist thought. Raskolnikov in *Crime and Punishment*, Ippolit and Myshkin in *The Idiot*, Shatov and Kirillov in *Demons*, as well as not just Ivan and Kolia Krasotkin but also Alyosha in *The Brothers Karamazov*, are all represented in ultimately positive terms despite their advocacy of at least some aspect of the nihilist package. Just as Dorothea, Will, and Lydgate are Eliot's heroes despite their mistakes, so these valiant if misguided young men are Dostoevsky's, if with the caveat that their various degrees of affinity for the nihilist/positivist project often lead to truly dire consequences. Even Lebeziatnikov for all his foolishness does a very good deed when he defends Sonia from Luzhin's trumped-up charge of theft; to his very real credit, the narrator explains, "he was a very kind little man" (*Crime* 365; *PSS* 6: 279). Certainly not every nihilist or nihilist-inflected character in Dostoevsky's novels displays similarly positive traits, and we might note Pyotr Verkhovensky in *Demons*, Rakitin in *The Brothers Karamazov*, or Luzhin himself, whose attempts to justify his utter and total selfishness offer a boiled-down version of nihilism. As Raskolnikov himself exclaims: "Get to the consequences of what you've just been preaching, and it will turn out that one can go around putting a knife in people" (151; *PSS* 6: 118). Many of his nihilists and nihilist-leaning characters are fundamentally good if confused young people, however, as Dostoevsky's message, like Eliot's, reflects an ambivalence derived at least in part from the real-life commitment to left-wing thought that played out a little differently for Dostoevsky and Eliot than it did for Hawthorne.

When Hawthorne departed Brook Farm after a stay of less than a year, he apparently lost the price of his investment in the enterprise, two shares at $500 each with an additional $500 thrown in. While he later sued to recover at least some of his money, it's not clear that he received anything back. Although, as Edwin Haviland Miller writes, "[t]his was a substantial sum on the part of a man who earned before the extras only $1500 annually" (E. Miller 189), still both Dostoevsky and Eliot paid a much steeper price for their embrace of the radical cause: while Dostoevsky's ten years of prison and exile need no explanation, the social ostracism that was Eliot's lot over the roughly two decades of her relationship with Lewes, including her family's cutting-off of all contact, was a source of deep and continuing pain. At the same time and in contrast to Hawthorne's brief flirtation, Eliot's and Dostoevsky's connection to left-wing thought never quite went away.

In Eliot's case, that continued attachment was reflected in the lifestyle choices that even an increasing advocacy of Tory values couldn't quite undo. Eliot in fact steered well clear of politics, to the point that, as Nancy Henry writes, even Eliot's "most political essay," the "Address to the Working Men by Felix Holt" (1868), "is in fact distinctly antipolitical" (Henry 142). To the disappointment of her readers then and now, Eliot also regularly refused to take any public and extra-literary stand on women's education or work opportunities. As she firmly explained to her close friend Barbara Bodichon, a leading advocate for women's rights, those sorts of statements "do not come well from me, and I never like to be quoted in any way on this subject" (143). As Henry summarizes, "[t]hese reservations about embracing even the limited political action that was available to her have led to the retrospective conclusion that she was 'conservative,'" a conclusion that Eliot herself seems to have supported when she told a friend that "the bent of [her] mind" was "conservative rather than destructive" (143–4). Despite Eliot's own claims, however, her early work on Comte, Feuerbach, and Strauss remained formative of her thought, and her unstinting support of Lewes's scientific work continued even after her partner's death. Above all, despite a yearning for respectability that was only partly assuaged by her late marriage to John Cross, a man some twenty-five years her junior, that she would even embark on a relationship with Lewes alongside her career as a writer speaks to a personal kind of radicalism that no public stance or lack thereof can entirely erase.

Dostoevsky, on the other hand, especially in his later years, was well-known for his vocal support not just for Russian Orthodoxy, but also Russian autocracy, Slavophilism, and anti-Semitism, all causes that would seem to place him firmly in the conservative camp; more recently, Russian president Vladimir Putin has cited Dostoevsky in defence of the invasion of Ukraine.[4] As Joseph Frank argues, however, any attempt at a strict classification fails to do justice to an idiosyncratic stance that never quite fit into any category, left or right, new or old, nihilist or not. Dostoevsky is in full-throated conservative cry in his *Diary of a Writer* (1876–7) when, as Frank writes, he "took up all the crucial social-political topics of the day" (J. Frank, *Mantle* 254). Even as the opinions that he expressed in the *Diary* often fell well to the right of the Russian mainstream, however, it was also exactly with that publication that Dostoevsky achieved remarkable popularity with the radical youth of his day. Frank accounts for this perhaps surprising effect by noting an ambiguity or openness in Dostoevsky's thought readily apparent to his earliest readers.

As Frank argues, Dostoevsky's anti-Semitic response to the "Jewish Question" is complicated by his response to a specific Jew, Arkady Kovner, who wrote a series of personal letters to Dostoevsky in protest. Dostoevsky not only responded by personal letter in turn but also published an article on the topic in the *Diary* itself, and if neither response sufficed either for Kovner or for readers today, what is striking is not just Kovner's continued engagement with Dostoevsky, but the terms in which he writes to him once again. "How can the Russian people not hate the Jews when its best representatives speak of them publicly as wild beasts?" he writes, "May you be forgiven, much-esteemed Feodor Mikhailovich, for this thoughtless paradox" (J. Frank, *Mantle* 319). Frank also draws our attention to an address that Dostoevsky wrote for the Slavic Benevolent Society in 1880.

This address was intended to be delivered to Alexander II on 19 February 1880 in honour of the twenty-fifth anniversary of his accession to the throne. The Society of which Dostoevsky was a member gave him the commission on 3 February, and on 5 February the terrorist group The People's Will set off a bomb in the Winter Palace just before a diplomatic dinner was set to begin. Although the tsar and his immediate guests were unharmed, ten soldiers on guard duty were killed and another fifty-six injured. On 19 February, Dostoevsky's text, already vetted by his colleagues at the Slavic Benevolent Society, was presented to the tsar. While Dostoevsky's address contains the requisite and, in his case, entirely sincere expression of his devotion to Alexander II, Dosto-evsky also takes the unusual step of referring directly to the forces that "'with unexampled audacity,' not long ago 'committed unheard-of evil deeds in our country, which caused shudders of outrage in our upright and mighty people and in the entire world'" (J. Frank, *Mantle* 481). As Frank notes, while it might seem inappropriate and possibly even risky for Dostoevsky as a former political prisoner to make any refer-ence to the assassination attempt at all, what is especially surprising is his description of the perpetrators of this recent offence as "'young Russian energies' whose motives, whatever their 'evil deeds,' could hardly be considered *entirely* criminal or wicked because they had been misguided in their *sincerity* and gone astray." In Frank's words, "[t]his repeated emphasis on the 'sincerity' of the radicals was hardly the lan-guage that the Tsar was accustomed to hear [*sic*] about those attempt-ing to destroy him and his régime" (481), and he quotes Alexander II's alleged response: "I never suspected the Slavic Benevolent Society of solidarity with the Nihilists" (482).

Although Alexander, if he uttered those words at all, could only have been speaking ironically, it is nonetheless the case that in a time of enor-mous polarization to the point of actual bomb-throwing, Dostoevsky was remarkable for his individual, personal engagement with represen-tatives of both sides, including not just the Jewish Kovner, but a range of interlocutors across the spectrum of Russian thought, from his twenty-three-year-old and accordingly left-leaning proofreader, V.V. Timofeeva, to Ober-Procurator of the Holy Synod, K.P. Pobedonosstev. As Frank argues, while Dostoevsky evidently possessed a remarkable ability to connect with his fellow human beings, his central political beliefs also transcended the ideological divides of his day. Where Hawthorne's var-ious political engagements, from his experiment at Brook Farm to his propagandistic work on behalf of then presidential candidate Franklin Pierce, his various political appointments and his scattered comments, including his ambiguous but vaguely pro-slavery comments in his 1862 essay "Chiefly about War Matters," stop at "higgledy-piggledy," Dos-toevsky's political views are made coherent by his love for and belief in the Russian people, an article of faith that he shared with both the Slavophile right and a new generation of left-wing populists. As D.N. Ovsianiko-Kulikovsky also writes, although "progressive young peo-ple, of course, could not accept his Slavophile point of view, his conclu-sions, and what we might call Dostoevsky's 'program,'" still his belief in "the lofty qualities of the Russian people and its great mission in the future renewal of mankind," the very belief on which the radical popu-lists based their own project to propagandize socialist ideas among the people, "was expressed by Dostoevsky with such deep faith, with such penetrating force of sincerity," that he served as an important source of inspiration. Indeed, Ovsianiko-Kulikovsky writes, in an echo of Alexander II, far from dissuading these young people from the revolutionary tendencies, his "preaching" served instead "to pour oil on the flames" (Ovsianiko-Kulikovskii 205). As significant as these real-life experiences and encounters surely were in shaping both Eliot's and Dostoevsky's attitudes to the new people and ideas of their day, still we can finally account for their affection for their nihilist and nihilist-tending heroes in other, more strictly literary terms. In both cases, their qualified but nonetheless enormous admiration for young people who want to change the world also reflects a close reading of Schiller.

Schiller as Poet, I: *L'Ami de l'humanité*

Like apparently everyone who was anyone in Russia in the first half of the nineteenth century, Dostoevsky first encountered Schiller by way of the Moscow production of *The Robbers* with the famous actor Pavel Mochalov as Karl Moor; at the time, Dostoevsky was all of ten years old.[5] Although Dostoevsky would cherish the memory of that production all his life, by his own account his real immersion in Schiller came during his years at the engineering school. As he protested in an 1840 letter to his brother Mikhail:

> You wrote to me, dear brother, that I have not read Schiller. You are wrong, my dear brother! I have learned Schiller by heart, I have talked in his language and have raved about him; and I think fate never did me a greater favor in all my life as when it allowed me to get to know the great poet at that period of my life; I could never have gotten to know him so well at any other time. (Kostka 216)

As Edmund Kostka explains, a few years later the two brothers even conceived the idea of publishing a complete translation of Schiller's works as "the road to financial success and literary glory" (217), a project that Dostoevsky abandoned only just before his life took its sudden turn in 1848. Well after his Siberian interlude and, indeed, to the very end of his life, Dostoevsky would continue to sing Schiller's praises, even while acknowledging his association with what had become for Dostoevsky a bankrupt political radicalism. As he wrote in *Diary of a Writer* in 1876:

> Even though the French Convention of 1793, when sending the certificate of citizenship *Au poète allemand Schiller, l'ami de l'humanité*, did perpetrate a

> beautiful, stately and prophetic act, nevertheless it did not suspect that at
> the other end of Europe, in barbarous Russia, that same Schiller was much
> more national and more akin to the barbarian Russians than to France –
> not only in those days but even later, throughout our whole century where
> Schiller, the French citizen and *l'ami de l'humanité*, was known, and then
> but slightly – only by professors of literature, and not even by all of them.
> Yet in Russia, together with Zhukovsky, he soaked into the Russian soul,
> left an impress upon it, and almost marked an epoch in the history of our
> development. (Kostka 234; *PSS* 23: 31)

Schiller here, even as adorned (twice) with the French epithet *l'ami de l'humanité*, is still the object of Dostoevsky's considerable veneration, not that the French epithet doesn't matter. *L'ami de l'humanité* points to Schiller not as he is, but as he has most often been read, as an advocate of a faux or at least entirely impractical idealism. This reading enjoys an especially rich history in Russia, where a certain class of freedom-loving young Russians not only celebrated Schiller, but even, astonishingly, attempted to act his plays out in real life.

In the conditions of an autocratic state, these attempts generally turned out badly, as in 1816 when the future poet Evgeny Baratynsky (1800–44) formed a "society of avengers" together with his classmates at the military academy. Allegedly inspired by *The Robbers*, the "avengers" went so far as to steal a gold snuff box containing five hundred rubles, only to find their allusion lost on their most important audience member, Alexander I; far from sympathizing with the ideal of the noble outlaw as exemplified by Karl Moor, he instead immediately reduced Baratynsky to the ranks. In another well-known example, Lotman accounts for Pyotr Chaadaev's otherwise inexplicable decision to retire from the army in 1820 again with reference to Schiller. In "The Decembrist in Everyday Life" (1984), Lotman argues that Chaadaev's intent in resigning immediately following an interview with the emperor was "to act out a variant of 'the Russian Marquis of Posa'" (Lotman 90), a tack that would entail persuading the autocrat to change his ways by first demonstrating his own personal incorruptibility. Unfortunately, Lotman writes, "Alexander I was a despot, but not one of the Schillerian persuasion" (92), and Chaadaev's would-be imitation of *Don Carlos* only "threw everyone into confusion" (88). As Martin Malia argues in *Alexander Herzen and Russian Socialism* (1969), the next generation of Russian reformers then picked up where the last one left off. In Malia's reading, as Herzen performed Schiller in emulation of the Decembrists, who were themselves engaged in an imitation of Schiller, we only withdraw ever farther from real life in a condition that Malia doesn't hesitate to characterize as adolescent.

For Malia, Schiller as "the poet of pure ego, of the limitless aggrandizement of self in fantasy," "of liberty stripped of all its political and social contingencies" (Malia 43), "presided over the greatest personal event" of Alexander Herzen's teenage years (45), his real-life friendship with Nikolai Ogarev. "The whole of Herzen's and Ogarev's youthful Decembrism," Malia writes, "soon came to be a fantasy adapted from *Don Carlos*" (49), a drive towards literary re-enactment especially marked in the two teenagers' decision to consecrate their dedication "to mankind and to Russia" in their famous pledge on the hills above Moscow. In Schiller's *Don Carlos*, Carlos's and Posa's similarly youthful commitment is mentioned only in passing, as Posa remembers "that oath / That in those wild, enthusiastic days / We swore together on the Host we shared"; it is now, he adds, for the adult Carlos to "make the vision true, / That daring vision of a brand new state / The godlike offspring of our friendship" (Schiller, *Plays* 259). In *My Past and Thoughts* (Былое и думы, 1870), Herzen dwells on the original moment at more length. As he recalls:

The sun was setting, the cupolas gleamed, the city spread out in a boundless expanse beneath the hill, a cool breeze blew over us; we stood for a long time, leaning together, and, suddenly, embracing, we swore, in the sight of all Moscow, to sacrifice our lives to our chosen struggle.... From that day forward the Sparrow Hills became a place of pilgrimage for us; we visited it once or twice a year, and always alone. (Malia 50)

Herzen at the time, of course, really was an adolescent, but adolescence in Malia's argument is also Schiller's permanent condition. "Schiller never grew up," Malia writes:

in the sense that he never experienced disillusionment with the ideals that he held at twenty. His last play, *William Tell*, has the same adolescent simplicity as his first, *The Robbers*, and all his heroes are exemplars of the simple, direct and monolithic self-assertion obliviousness of the external world, which is the special illusion of adolescence. (42)

This Schiller, Schiller *l'ami de l'humanité*, is finally on full display in Dostoevsky's novels.

As Robert Belknap writes, "Dostoevsky's characters talk more about Schiller's works than about any other writings except Pushkin's, Gogol's, possibly Shakespeare's and the Bible" (Belknap, *Structure* 28).[6]

These allusions are also overwhelmingly negative, so much so that the hint of mockery that we find in *Diary of a Writer* regularly expands in the novels into full-blown parody. The narrator's repeated evocation of "the beautiful and the lofty" in *Notes from Underground* only points to the jarring disconnect between his high ideals and his sordid reality, while General Ivolgin's entirely manufactured claim to have nearly fought a duel with Prince Myshkin's father "[a]cross a handkerchief" (*Idiot* 95; *PSS* 8: 81) in *The Idiot* betrays his reading of Schiller's *Cabal and Love* (*Kabale und Liebe*, 1784), as does his "rapturous" (*Idiot* 123; *PSS* 8: 104) conversation generally; Ivolgin may be a liar, a sponger, and a drunk, but as Lebedev explains, "he's a man of the loftiest feelings" (446; *PSS* 8: 373). In *Crime and Punishment*, Schiller's very name is offered up as a pejorative term. It is first Raskolnikov who sneers at his family's sacrifice and their "beautiful, Schilleresque souls" (*Crime* 42; *PSS* 6: 37), then Porfiry Petrovich who asks, "Why are you smiling again – because I'm such a Schiller?" (460; *PSS* 6: 352). All the while, the real "Schiller" is the philanthropic murderer himself, as Svidrigailov obligingly points out. "Look at our Schiller, what a Schiller, just look at him!" (482; *PSS* 6: 371), he mocks Raskolnikov. "Schiller is constantly being embarrassed in you" (484; *PSS* 6: 373). This negative invocation of Schiller reaches its height in *The Brothers Karamazov*.

In *The Brothers Karamazov* the father Fyodor Pavlovich presents the threat that his sons pose to him exactly in terms of *The Robbers*. "This is my most respectful Karl Moor, so to speak," Fyodor Pavlovich says, "and this son, the one who just came in, Dmitri Fyodorovich … is the most disrespectful Franz Moor, … and I, I myself in that case am the *regierender* Graf von Moor!" (*Brothers* 71; *PSS* 14: 66). As damning as Fyodor Pavlovich's reading may be, troubling, too, is the attempt of both the prosecution *and* the defence to make use of Dmitri's "love for Schiller, his love for the 'beautiful and lofty'" (743; *PSS* 15: 169), although it is certainly true that Dmitri quotes from Schiller's poetry at length, as does Ivan more briefly. If Ivan's reference to Schiller in its very understatement might seem the most sincere, it is also the most unexpected, as the narrator, Alyosha, and even Mme. Khokhlakov all note. "And did you notice," she asks Alyosha, "what youthfulness came out in Ivan Fyodorovich just now, he said it all and walked out! I thought he was such a scholar, an academician, and he suddenly spoke so ardently – ardently, openly, and youthfully, naively and youthfully, and it was all so beautiful, beautiful, just like you…. And he recited that little German verse, just like you!" (195; *PSS* 14: 178). As "Schiller" now points to murder of various sorts, not to mention to both Fyodor Pavlovich's and Mme. Khokhlakov's particularly cheap and performative versions of radical thought, it might seem that Dostoevsky also sees Schiller as adolescent, exactly as in his 1875 novel of the same name, perhaps meaningful to him when he was an adolescent himself, but by *The Brothers Karamazov* well outgrown. Parody is a complicated literary operation in Dostoevsky, however, and here reflects his recognition that what his beloved Schiller has offered all along is not an answer, but a rich and productive problem.

Bad readings, after all, are everywhere in Dostoevsky, not that his intent is that we will follow suit, and if Malia's Schiller, the silly and at the same time dangerous one, is the one that Dostoevsky's characters know best, Dostoevsky himself had access to other readings. There is no doubt that Schiller as poet and especially as playwright, not just in Russia, but across Europe, has most often been read exactly as the French Convention would have him, as a call to revolution. Even so, Schiller as philosopher has long been appreciated in entirely different terms. This Schiller is not the champion of an insistent, idealized, and even illusory freedom, but instead an advocate of exactly the alternative materialism that this work has explored, with all the emphasis on uncertainty, contingency, and complicated interrelationships in the conditions of the real world that that materialism entails.

Schiller as New Materialist Philosopher

In *The Discovery of Chance*, Aileen Kelly argues that Herzen himself made a sharp pivot from the first Schiller to the second as his more mature scientifico-political stance took shape.[7] As early as 1840, she writes, Herzen "turned his back on Schiller as the poet of adolescent dreams," only to reverse course once more. "I soon came to my senses," Herzen writes, "blushed at my ingratitude, and with burning tears of repentance threw myself into [his] embrace" (Kelly 221). According to Kelly, Herzen's focus had now shifted from Schiller's plays to his essays, including "On Grace and Dignity" ("Über Anmut und Würde," 1793) and *Letters on the Aesthetic Education of Man* (*Über die ästhetische Erziehung des Menschen in einer Reihe von Briefen*, 1795). More importantly, he was now reading Schiller through the lens of yet another new discovery: Feuerbach. Lest Schiller and Feuerbach strike us, in her words, as "strange bedfellows," Kelly makes the case for their affinity with brief reference to Schiller's own background in the sciences.

In contrast to Dostoevsky, Schiller's undergraduate training was in medicine, not engineering, and his less than happy stint in the military academy from 1775 to 1780 came at the insistence not of his father, but of the local ruler, Karl Eugen, Duke of Württemburg. The parallels are nonetheless striking, including in terms of the impact of this early work on his later writing. Although Schiller, again like Dostoevsky, left the sciences for a career in letters shortly after graduation, Kenneth Dewhurst and Nigel Reeves, like Kelly, argue the "significance of the young Schiller's experiences and intellectual training as a doctor and psychologist for the literary, philosophical and aesthetic works on which his fame rests" (Dewhurst 307). Schiller in fact wrote three dissertations in pursuit of his degree, the first and the third, "The Philosophy of Physiology" ("Philosophie der Physiologie," 1779) and the "Essay on the Connection between the Animal and the Spiritual Nature of Man" ("Versuch über den Zusammenhang der thierischen Natur des Menschen mit seiner geistigen," 1780), attesting to his abiding interest in what Kelly calls "the empirical approach to the relation between body and mind then being pioneered by such figures as the Swiss physiologist Albrecht von Haller" (Kelly 222).[8] Where Kelly in the Russian context makes only a brief allusion to this more materialistically and even scientifically inclined Schiller, both Jane Bennett and John Tresch explore this other Schiller at more length, although with reference not to his early training and affinity for Feuerbach, but in contrast to Immanuel Kant (1724–1804).

Fleeting mention of Kant has already appeared in these pages, always with reference to an advocacy of mind over matter, and Bennett and Tresch take Kant as their point of departure in presenting Schiller as a New Materialist ally of sorts. In *The Enchantment of Modern Life* (2001), Bennett is the more measured of the two, as she argues that Schiller's aesthetics and especially his theory of the "play drive" offer only a qualified corrective to the abstractions of Kantian morality. Schiller's "play-drive," Bennett explains, mediates between his "sense-" and "form-drive" as "a protean force that can be cultivated into an aesthetic disposition, a state of being wherein both our active and passive natures are engaged without either one dominating" (Bennett, *Enchantment* 138). Bennett remains unconvinced by Schiller's belief in the power of a "proper aesthetic education" to "harmonize or reconcile" (140). Even so, she sees his view of the "ethical insufficiency of intellect" and especially his attempt "to assign a pointedly ethical role to bodily affect" (140) as a small but significant step towards the model of ethics that her "enchanted materialism" sets in motion. While he follows Kant in positing a transcendental Reason, she writes, Schiller "affirms, more actively than Kant, the need for Reason to employ a phenomenal champion" (138), and she quotes approvingly from his *On Aesthetic Education*:

Reason has accomplished all that he can accomplish by discovering the law and establishing it. Its execution demands a resolute will and ardour

of feeling. If Truth is to be victorious ... she must ... appoint some drive
to be her champion in the realm of phenomena; for drives are the only
motive forces in the sensible world. (139)

Tresch brings Schiller still closer to the New Materialist camp when
he casts Schiller's philosophical grounding in the material conditions
of the real world in terms of his impact on yet another major figure in
nineteenth-century science: Alexander von Humboldt (1769–1859).

In the chapter in *The Romantic Machine* devoted to Humboldt,
Tresch argues that Schiller's aesthetics found concrete expression in
Humboldt's science and especially in his use of a wide array of scientific instruments. Tresch gives a wonderful account of Humboldt's
reliance on:

> an embarrassment of devices for registering a huge range of phenomena:
> chronometers, telescopes, quadrants, sextants, repeating circles, dip nee-
> dles, magnetic compasses, thermometers, hygrometers (based on a strand
> of human hair that grew longer or shorter depending on the moisture
> of the air), barometers, electrometers, eudiometers (to measure the air's
> chemical composition). (Tresch 77)

Central to Tresch's argument is Humboldt's sense of his tools as entities in their own right endowed with their own agency. "He typically
identified [his instruments] by the patronymic of their makers," Tresch
explains, "and went to great lengths to assure their well-being; his most
cherished compasses, barometers, and sextants were discussed with the
same enthusiastic affection as his dearest friends" (78). In this insistence on the "mutual respect among users, tools, and their objects" (76),
what Tresch also calls a "polity of scientific instruments and users" (85),
Tresch marks the move away from Kantian abstraction that he again
associates with Schiller.

"Like Schiller's 'fine art,'" Tresch writes, "Humboldt's instruments
were the concrete media occupying the milieu, the 'halfway-place'
between the mind and the world: the concrete locus for the fusion of
sense and intellect" (Tresch 78). "Humboldt's work with instruments,"
he adds, "showed that in the same moment that perception was somatized and particularized – thereby dismantling the Kantian notion that
the categories of experience were universal properties of the transcendental ego – these categories were *made into universals in practice*, in the
technical apparatus of the observatory sciences" (78). The "fusion of
sense and intellect" that Tresch, like Kelly, explicitly associates with a
certain strain of science, would seem utterly at odds with the idealistic

and even adolescent invocation of individual liberty that Malia finds in Schiller's plays. The ambivalence that Dostoevsky expresses as he praises the very writer whom he also mocks as *"l'ami de l'humanité"* isn't just a reflection of a perceived difference between Schiller as poet and Schiller as philosopher, however, nor even of the wide gulf between Schilleresque ideals and Russian reality. Dostoevsky's pointedly mixed message most importantly argues for a more complete reading of Schiller as poet in the first place.

Although students of the Russian nineteenth century lay emphasis on what they take to be Schiller's appeal to abstract idealism, equally striking is Schiller's reliance on a "sensation" of his own. As Dewhurst and Reeves note, it is not just that the characters in his plays are often remarkably physically involved – in *The Robbers*, Franz "paces violently" (Schiller, *Robbers* 255), faints, and even "writhes ... in fearful convulsions" (287) – but that his audience seems to have responded in kind. As an eyewitness account famously recalled the chaos that ensued when *The Robbers* first premiered: "[t]he theatre was like a madhouse: rolling eyeballs, clenched fists, stamping feet, hoarse cries in the auditorium! Complete strangers fell sobbing into each other's arms, women staggered almost fainting to the door" (Reed 21–2). Russian critic Vissarion Belinsky (1811–48), another young audience member at the Moscow production, remembered *The Robbers* as "that wild, flaming dithyramb erupting like lava from the depths of a young, dynamic soul" (Kostka 88). On just reading the same play in 1794, an excited young Coleridge wrote to his friend Robert Southey: "My God! Southey! Who is this Schiller? This Convulser of the Heart?" (Schiller, *Don Carlos* ix). For all its apparently extraordinary effects, "sensation," as Schiller would have it as early as 1781, is once again not a matter of elevating bodies over minds or the other way around. The sad tale that his plays almost invariably tell is instead of an unfortunate, because entirely avoidable, conflict between the two.[9] It may be that youth inevitably grows old and lofty ideals find no bearing in the uncertain and always-contingent conditions of the real world. It is nonetheless their tragedy that Karl Moor in *The Robbers*, as well as both Posa and Carlos in *Don Carlos*, turn away from the "somatized and particularized" interactions with real people in the world that would temper their finally deadly abstraction.

Schiller as Poet, II: The Problem of Ardour

In *The Robbers*, Karl, also known as Robber Moor, is the hero of the piece, the handsome eldest son of the Count von Moor and beloved of the lovely Amalia. What starts as a youthful kicking-over-of-the-traces escalates into actual crime when Karl finds his way home barred by his scheming younger brother Franz. It is only when Franz falsely informs him that his father's forgiveness will not be forthcoming that Karl recasts himself as the leader of a robber band, and even then he is not entirely committed to Malia's "liberty stripped of all its political and social contingencies" (Malia 43). When his robber band rescues their comrade Roller, for example, Karl is dismayed to learn that along the way his men have killed women and children, the sick and the old. "Roller," he says, "your life is dearly bought" (Schiller, *Robbers* 231). By the play's end, Franz has left their imprisoned father to die, and Karl can only fulfil his obligation to his band of merry men at the cost of Amalia's life, not that that sacrifice offers anything in the way of a resolution. With the cry "Moor's love shall die by Moor's hand alone" (295), Karl kills Amalia, only to turn himself in to the authorities. "Oh, fool that I was," he says, "to suppose that I could make the world a fairer place through terror, and uphold the cause of justice through lawlessness" (296). Although Franz by this point has committed suicide by self-strangulation, still the victory is his: like Franz, Karl ends by choosing death over "living life."

Franz is the real problem as, while keenly aware of the workings of bodies, he tries always to suppress that instinct. Despite his own "convulsions" and "furious" rushing about, despite his vivid descriptions of Karl's supposedly rotting flesh and the physical effects of terror (Schiller, *Robbers* 210), Franz sees himself as "this dry, cold, wooden Franz" (186) with a grudge against nature. "Why was I not the first to creep out of our mother's womb?" he asks, "Why not the only one? Why did nature burden me with this ugliness?" (189). If it is "natural" to feel an obligation to one's brother and his father, Franz will have none of it. "I have heard a great deal of idle talk about something called love of one's kin," Franz says. "Just consider this extraordinary conclusion, this ridiculous argument from the proximity of bodies to the harmony of minds ... from the uniformity of diet to the uniformity of inclinations" (190). Karl, at least in his brother's account, starts the play at home in his body and in nature, as Franz recalls for his father's benefit:

> The fiery spirit that burns in the lad ... that makes him yearn so keenly for every kind of beauty and grandeur; the frankness that mirrors his soul in his eyes, the tender feeling that melts him to tears of sympathy at any sight of suffering, the manly courage that sends him climbing hundred-year-old oak trees and leaping ditches and fences and foaming rivers. (185)

The effect of Franz's scheming, however, is to empty Karl of his natural inclinations. When we first meet Karl, he is still capable of casting his rebellion as natural in its own way. "There they go," he says disparagingly, "smothering healthy nature with their ridiculous conventions" (192). By the play's end, however, Karl is no longer a force of nature, but stands outside it, as he puts it, "at the limit of a life of horror": "*two men such as I,*" he says, "*would destroy the whole moral order of creation*" (296). As Karl exercises his "one last merit, of dying for justice of [his] own free will" (296), he remains the noble figure that he has been all along. In choosing death over life, however, he nonetheless makes a fundamental mistake, exactly the same mistake that leads to the tragic conclusion of *Don Carlos*.

In *Don Carlos*, the Marquis of Posa, like Karl, tries and fails to realize his nobility at the expense of human relationships, and above all his relationship with Carlos, son and heir to the King of Spain. In his later *Letters on Don Carlos* (*Briefe über Don Carlos*, 1788), Schiller is at pains to argue that any real friendship was only ever on Carlos's side, as Posa, ever the "citizen of the world" (Schiller, *Plays* 320), simply doesn't operate in those terms. While "[l]ofty, effective benevolence toward the whole in no way excludes a loving interest in the joys and sorrows of an individual" (321), Schiller writes, the priority that he gives to the many nonetheless leads Posa to view Carlos as "the *single indispensable tool*" for his "one ardently and resolutely pursued goal" (323), the liberation of Flanders. Posa arrives at the court with the idea that Carlos should propose himself as governor of Flanders, only to find Carlos absorbed in his hopeless love for his stepmother, the queen. The quick-thinking Posa promptly engineers an unsanctioned meeting between the two, with the happy result that the virtuous Elizabeth herself urges Carlos to take up the apparently greater cause. When the king then rejects Carlos's proposal out of hand, Posa changes tacks once more and adopts the role of the king's friend and confidant. Always with the higher aim of saving Flanders, Posa goes so far as to pretend to betray Carlos, handing over a selection of his letters to the king and even having him arrested. Unfortunately, this complicated attempt to bend real affections in the world to his own lofty aims only ends in his own and Carlos's undoing.

As Posa admits to Carlos at the start of Act V, "[t]he structure that I built / Comes tumbling down – for I forgot your heart" (Schiller, *Plays* 270). Despite Posa's clarity at that moment, Lesley Sharpe argues that the unhappy results that ensue are not his fault, nor even entirely unhappy. While the immediate consequences for Flanders are grim, Sharpe reminds us that our own knowledge of history "mitigates that tragedy." He also sees the "true tragedy" as "the destruction by the King himself of his own family as prelude to his attempted destruction of the Netherlands" (Sharpe 95). Undoubtedly, the king is a large part of the problem, as his handing over of his own son to the Inquisition is pointedly presented as again unnatural. When Philipp worries, "I offend / The voice of Nature – can you also silence this mighty voice?" the Grand Inquisitor is firm: "No voice of Nature is / Of any worth before our Faith" (Schiller, *Plays* 299). As Schiller himself argues in his later *Letters*, however, the crime against nature is first Posa's, not just in his trifling with Carlos's affections – "What is the Queen to *you*?" Carlos asks. "Did *you* love her? / And should your austere virtue answer to / The unimportant worries of my love?" (270) – but in his insistent focus on the whole at the expense of the part. In a moment of "intense emotion," Posa takes leave of Elizabeth with the words "– O God! But life is beautiful!" (263). Unfortunately, Schiller writes in the *Letters*, "he makes this discovery too late" (346). As Schiller also explains, "the most unselfish, purest, and noblest person very often, because of enthusiastic devotion to *his idea* of virtue and happiness to be gained, is very often displayed dealing just as arbitrarily with individuals as even the most selfish despot, because the object of both their endeavors lies *within* them, not *outside* of them" (339). Posa is a little too like the Grand Inquisitor when the latter reminds Philipp that for him men should be "numbers, nothing more" (297). By the play's end, he has also convinced Carlos of the same.

With Posa already dead, Carlos is finally ready to take up the cause of Flanders; as he tells the queen in the play's final scene, his passion for her now "dwells in graves / Where dead men die" (Schiller, *Plays* 301). Like Franz before him, he then pledges to take up arms against his father. "The voice of Nature," Carlos claims, "Has totally departed from my breast" (302). In the absence in both instances of what Schiller in the *Letters* calls "*love* for a *real object*" (339), Carlos, like Posa, chooses death, and death promptly arrives. As the king steps between Carlos and the queen, the queen collapses, her death evidently his work. Philipp then hands Carlos over to the Grand Inquisitor with the play's last line: "So, Cardinal, / Now I have done my part. See you do yours!" (303). Elizabeth is already dead, although not actually at the hands of her beloved; Posa is dead, too, shot on order of the king. Only Carlos remains, but not for long. With his final words, we understand, Philipp has condemned his son to death, as *Don Carlos*, like *The Robbers*, offers no means of reconciling the competing demands of ideal and real, mind, body, and material world on the diegetic level. That reconciliation instead operates only on the level of audience, not just when we take the long view that the history of the Netherlands offers, but when Schiller deliberately evokes our clenched fists and stamping feet and even our attempts at real-life re-enactment. Few writers enjoy such an extreme version of reader response, and Dostoevsky didn't, not quite. Where Dostoevsky adds to Schiller's expression of minds and bodies in the world, however, is on the level of plot.

The Schiller that runs all through Dostoevsky's novels, including in his own rendering of the Grand Inquisitor, is perhaps most often a reference to adolescent dreams. "Look at our Schiller, what a Schiller, just look at him!" Svidrigailov says (*Crime* 482; *PSS* 6: 371). Even when Dostoevsky is at his most mocking, however, "Schiller" always evokes the same tragedy. In Dostoevsky, as in Schiller, ardour cut free from any real substance can produce a kind of despair in the souls of impressionable adolescents, including not just a "pensive" (*Demons* 41; *PSS* 10: 35) young Stavrogin, but also "bookish, dreamy, self-enclosed, and fantastical" Nastasya Filippovna (*Idiot* 569; *PSS* 8: 473). The commitment to making "bookish" ideals real can also quickly devolve into a kind of despotism. Either way, the result is death, although in sharp contrast to the deaths of Amalia and Karl, Elizabeth, Posa, and Carlos, murder/suicide in Dostoevsky isn't an ending, not just for his readers, but even for his characters. Where tragedies by definition end with death, after all, most novels don't. Novels cast as murder mysteries go so far as to take death as the premise for the plot's unfolding, and in both *Crime and Punishment* and *The Brothers Karamazov*, death is explicitly the start of a new story.

The most obviously ardent of the Karamazov brothers is Dmitri, not just in his lengthy quotes from Schiller at his most exalted, but also in the titles of the three chapters that introduce us to him: "The Confession of an Ardent Heart. In Verse" ("ИСПОВЕДЬ ГОРЯЧЕГО СЕРДЦА. В СТИХАХ"), "The Confession of an Ardent Heart. In Anecdotes" ("ИСПОВЕДЬ ГОРЯЧЕГО СЕРДЦА. В АНЕКДОТАХ"), and "The Confession of an Ardent Heart. 'Heels Up'" ("ИСПОВЕДЬ ГОРЯЧЕГО СЕРДЦА. 'ВВЕРХ ПЯТАМИ'"). Ivan's less public but equally fundamental indulgence in Schiller is nonetheless evident not just in his unexpected quotation from Schiller's ballad "The Glove" ("Der Handschuh," 1797), but above all in his lifting from *Don Carlos* for his own "Grand Inquisitor." Indeed, for all Dmitri's play at the kind of "honour" that Schiller's heroes also display, it is Ivan who most despairs at the discrepancy between the world as he wants it to be and the world as it is, just as it is Ivan who is most deeply implicated in the deaths of their father and (half-)brother.

In the midst of a large number of real brothers and fathers, not just those of various biological degrees, but also the monks and elders at the nearby monastery, Ivan's love, like his mathematics, has no bearing on the real world only because, like Karl Moor, he won't let it. Even while acknowledging his love for "the sticky little leaves that come out in spring," the "blue sky" and even "some people ... whom one loves sometimes, would you believe it, without even knowing why" (*Brothers* 230; *PSS* 14: 210), Ivan, like Mme. Khokhlakov, insists that "[i]t's still possible to love one's neighbor abstractly, and even occasionally from a distance, but hardly ever up close" (237; *PSS* 14: 216). In the many examples that the novel offers of ardour gone awry, from the play-acting of Mme. Khokhlakov's sentimentality to the despotic control that Kolia Krasotkin exerts over the neighbourhood children, we see reflections of the often dire consequences that result in the absence of "*love* for a *real object*" (Schiller, *Plays* 339). Ivan's father and brother die because, like Posa, Ivan has decided that "everything is permitted," even if Ivan himself never quite says those words. In *Crime and Punishment*, in contrast, Raskolnikov comes a little closer to speaking the problem of ardour. His example also makes more explicit the way back to "living life."

We hear snatches of higher ideals in the conversation that Raskolnikov happens to overhear in a tavern when another "ardent" [загорячился] (*Crime* 65; *PSS* 6: 54) student, not Raskolnikov, unfolds a plan for murdering the same pawnbroker:

> Kill her and take her money, so that afterwards with its help you can devote yourself to the service of all mankind and the common cause: what do you think, wouldn't thousands of good deeds make up for one tiny little crime? For one life, thousands of lives saved from decay and corruption. One death for hundreds of lives – it's simple arithmetic! (65; *PSS* 6: 54)

It's a vision of the greatest good for the greatest number, a rational utopia based on the same sort of logical, scientific thinking that Raskolnikov himself uses when he describes his own "calculations" for the crime as "true as arithmetic" (59; *PSS* 6: 50). This particular kind of giving turns out to be taking, however, as philanthropy and cold logic dissolve into an aspiration for power and Raskolnikov himself ultimately admits that he killed two defenceless women only to answer the question: "Would I be able to step over or not! Would I dare to reach down and take, or not? Am I a trembling creature or do I have the *right*" (419; *PSS* 5:3 22). As he realizes in the novel's final pages, the answer is no. In striking contrast to Schiller's Karl, Carlos, and Posa, however, Raskolnikov learns that lesson without dying himself.

The very last lines in *Crime and Punishment* read:

> But here begins a new account, the account of a man's gradual renewal, the account of his gradual regeneration, his gradual transition from one world to another, his acquaintance with a new, hitherto completely unknown reality. It might make the subject of a new story – but our present story is ended. (*Crime* 551; *PSS* 6: 422)

Like Ivan apparently on his way to recovering from his bout of brain fever as *The Brothers Karamazov* concludes, Raskolnikov finds new life as cast in explicitly Christian terms. As chapter three has already argued, this new life is at the same time and as part of the same whole the response of their bodies to acts undertaken at the urging of their rational minds alone. Dostoevsky's repeated allusions to Schiller add one last facet to a "living life" that in politics, as in science, is always both one and many, as the return of Dostoevsky's heroes to the land of the living is finally also made possible by their embrace of even just one real other in the world.

When we learn in the last paragraphs of *Crime and Punishment* that Raskolnikov keeps a copy of the Gospels under his pillow, we are also told that he has never looked inside, and he doesn't do so now. As the narrator explains, Raskolnikov's epiphany is instead that he loves Sonia as she loves him: "They were resurrected by love; the heart of each held infinite sources of life for the heart of the other" (*Crime* 549; *PSS* 6: 421). It is only after this realization that Raskolnikov's thoughts turn to the Gospels, and only because his copy of the Gospels is actually hers. One thought flashes through his mind: "Can her convictions not be my convictions now? Her feelings, her aspirations, at least …" (550, ellipsis in the original; *PSS* 6: 422). Certainly, Raskolnikov seems to have found the Christian God as he rediscovers the sensations of his own body. Still, what we are told in so many words is that Raskolnikov reaches the end of his journey when he finds an other who is also a self.

When scholars of Dostoevsky understand Raskolnikov's achievement in terms other than those offered by Russian Orthodoxy, we often recur to Bakhtin's notion of dialogism. For all that it adds to our understanding of Dostoevsky, however, dialogism is a concept that pertains more to language and to the interrelationship of words on the page than it does to politics as it plays out among real bodies in the world. The human relationships that conclude Eliot's *Middlemarch*, on the other hand, readily lend themselves to a political meaning alone, although one that many readers find less than satisfactory: as Fred and Mary, Dorothea and Will join in the conventional bonds of matrimony, it may seem that the once radical Miss Evans has thrown her lot in with bourgeois society once and for all. To read Dostoevsky and Eliot not just once more against one another but now also together with Schiller, however, is to argue one last time for Tresch's "halfway-place." In the ideal world that Schiller's heroes imagine, there is no place for "living life" and no political way forward. In direct contrast, Dostoevsky's and Eliot's idealistic young heroes, at least some of them, live to learn better, as the sometimes adolescent but always ardent enthusiasm that drives left-wing, socialist utopian politics meets up with real minds and bodies in the world and a different political formation emerges: fellowship.

The Politics of Fellowship

"Fellowship" is Auerbach's term in "Dorothea's Lost Dog" (2006) when she writes that Dorothea in refusing the gift of a little dog in the very first pages of the novel "begins her story by spurning the greatest prize the secular world of Middlemarch holds: the treasure of fellowship" (Auerbach 88). What Dorothea has instead is her unquenchable and, as Auerbach notes, sometimes unbearable ardour. Ardour as cast in entirely positive terms is Dorothea's most striking attribute throughout the novel, with her "exalted enthusiasm" (Eliot, *Middlemarch* 28), her yearning for "action at once rational and ardent" (86), and the "ardent character" (772) that leads her to reach out to Lydgate in his time of need. As the narrator entirely approvingly describes Dorothea at that moment: "The presence of a noble nature, generous in its wishes, ardent in its charity, changes the lights for us" (762). The same quality lifts Eliot's leading men in the same way. Will is a notably ardent fellow, not least when he addresses the still-married Dorothea "ardently," the two "looking at each other like two fond children who were talking confidentially of birds" (392), as is Lydgate, whose "intellectual ardour" is both cause and effect of the true and truly beautiful scientific vocation that appeared when the very young Lydgate read the encyclopedia entry for "Anatomy" and "the world was made new to him by a presentiment of endless processes filling the vast spaces planked out of his sight by that wordy ignorance which he had supposed to be knowledge" (144). If the fears engendered by what Mr. Brooke calls Will's "enthusiasm for liberty, freedom, emancipation" (359) come to naught, still the murder that is entirely of Mr. Hawley's imagining in his case is a little more real in the case of Eliot's other two heroes, Dorothea with her implication in the death of her husband,

and Lydgate's in the death of Raffles. Ardour also leads Dorothea into a confusion of giving with taking that extends beyond Casaubon and into a general way of being.

As again Celia explains, Dorothea "likes giving up" (Eliot, *Middlemarch* 18), including not just her late mother's jewelry and the riding that she admits she enjoys "in a pagan sensuous way" (10), but even, as Auerbach notes, the little dog that her neighbour and sometime suitor Sir James Chettam tries to present as a gift. As Auerbach argues, while Dorothea truly is a noble, self-sacrificing sort, it is also her habit to make her friends and neighbours, and especially her sister, feel that superiority in a small but insistent bid for power. As much as Celia admires Dorothea, she also chafes at her control, as the narrator remarks in the wake of the jewelry episode. "Since they could remember," the narrator explains, "there had been a mixture of criticism and awe in the attitude of Celia's mind towards her elder sister. The younger had always worn a yoke; but is there any yoked creature without its private opinions?" (15). Dorothea's refusal of both jewelry and dog may seem relatively harmless in comparison to her really very unfortunate treatment of Casaubon, not to mention, on the extreme, Russian end of things, Raskolnikov's double murder. Even so, as Auerbach insists, Dorothea's "rejection of animals is not only a denial of physicality and the body, though it is that; it also withholds the sympathy so needed in Middlemarch, yet so generally absent from it" (90). Although Dorothea's friends don't cry out "Schiller" at this moment, they certainly could: Eliot's starting point is the same.

In contrast to Dostoevsky, Eliot didn't encounter Schiller until she was all of twenty years old. In a characteristic expression of her own "intellectual ardour," however, once she did, she went straight to the source. Eliot turned to the study of German in March 1840, six months later she was already reading Schiller's *Maria Stuart* (1800), and over the course of the next five years, she read all the plays and a great deal of the poetry along with the *History of the Thirty Years' War* (*Geschichte des dreißigjährigen Kriegs*, 1790) and Carlyle's *Life of Schiller* (1825). Her response, like Dostoevsky's, was immediate and intense. As her friend Mary Sibree later recalled, "Placing together one day the works of Schiller, which were in two or three volumes, Miss Evans said, 'Oh, if I had given these to the world, how happy I should be!'" (Guth 25). Eliot only occasionally makes explicit reference to Schiller in her own fiction, in *Daniel Deronda* (1876) when Mirah compares Gwendolyn to the Princess Eboli, for example, or when the young Mab Meyrick exclaims, "I want to do something good, something grand.... It makes me like Schiller – I want to take the world in my arms and kiss it" (Eliot, *Daniel*

Deronda 168). Deborah Guth argues that Eliot's reading of Schiller is largely reflected instead in a tendency to deploy extended metaphors. While chapter four has already explored Eliot's figurative language in other terms, there is a distinct echo of Schiller as Tresch describes him in Guth's explanation that, because Eliot's metaphors are "[s]ituated half way between the abstract and the concrete," they "'represent' abstractions in physical form, making 'tangible what analytically remains ungraspable', while at the same time ... 'provid[ing] ... access to what is beyond empiricism'" (Guth 11). As Guth notes, the plots that Eliot unfolds also unfailingly revolve around the problem of ardour.

In *Middlemarch*, Eliot's ardent young people are never quite as extreme as Dostoevsky's, and their struggles not quite as desperate. In the absence of *"love* for a *real object*," however, the confusion of giving with taking plagues almost all the instances of philanthropy in Eliot as in Dostoevsky, a point made especially clear in the practice of an older generation. Just as Luzhin in *Crime and Punishment* casts Raskolnikov's would-be philanthropy in the most unflattering of lights, so Casaubon, Peter Featherstone, and Bulstrode with their more or less open attempts to use giving as a source of power all comment on Dorothea's earnest efforts to do good. The intensely evangelical Bulstrode is the most obvious in this regard, as he excuses what was in effect the theft of money itself illegally obtained by the greater good that his crime might serve. "Those misdeeds, even when committed – had they not been half sanctified by the singleness of his desire to devote himself and all he possessed to the furtherance of the divine scheme?" (Eliot, *Middlemarch* 525). As Dorothea also discovers, true giving is instead free of the expression of power and marked by a corresponding ability to receive. Like Dostoevsky's young heroes, Dorothea with the "ardent nature" that "turned all her small allowance of knowledge into principles" (193) starts the novel too "theoretic" (28), and she quickly finds her way to her disastrous first marriage and her husband's death. She is given a second chance, however, when she learns what it is to truly give, first to Lydgate, then to Rosamund, and finally to Will.

Dorothea's gift to both men takes the form of money. To Lydgate she gives the money that he so desperately needs to repay his debt to Bulstrode, while Will accepts her personal fortune as well as the giving-up of Casaubon's money that marriage to Will entails. Lest we be tempted to read the latter encounter in other, less mercenary terms, we should note that it is Dorothea who proposes marriage to Will and that she does so entirely with reference to financial transaction. As she says "in a sobbing, childlike way": "We could live quite well on my own fortune – it is too much – seven hundred-a-year – I want so little – no new clothes – and I will learn what everything costs" (Eliot, *Middlemarch* 812). What allows both men to accept Dorothea's money is the "direct fellow-feeling with individual fellow-men" that both understand in terms of the institution of marriage as nineteenth-century English society conventionally would have it. Where Lydgate, of course, only wishes that he could marry Dorothea, Will is fortunate enough to actually do so, and by the "Finale" he is "an ardent public man, working well in those times when reforms were begun with a young hopefulness of immediate good which has been much checked in our days, and getting at last returned to Parliament by a constituency who paid his expenses" (836). Despite the ambivalence of "young hopefulness," it's a happy ending with Will's ardour intact, although Dorothea's friends and even some of her readers wonder if Dorothea herself has done quite as well out of the exchange.

A real-world dialogicity would seem exactly what Will has on offer as "a creature who entered into every one's feelings, and could take the pressure of their thought instead of urging his own with iron resistance" (Eliot, *Middlemarch* 496). In the "Finale," however, we learn that Dorothea is now "only known in a certain circle as a wife and mother" (836) in an apparently total dissolution of self that might seem to exceed and so even undermine the requirements of fellowship. As the narrator acknowledges, "[m]any who knew her thought it a pity that so substantive and rare a creature should have been absorbed into the life of another" (836). Many of her readers have felt the same. Marian Evans, we should recall, not only cohabited for many years with an already married man, but went on to achieve international fame as a writer while also, according to one close friend, practising birth control; after Lewes's death in 1878, she shocked her admirers all over again when she married John Cross, a man more than twenty years her junior. As George Eliot, she finally operated in an environment increasingly vocal in raising what the Russians called the "Woman Question." Given Eliot's own life experiences, Dorothea's readers, like her friends, might well expect that Eliot herself would answer that question in sharper terms. Fellowship in literature as in life, however, steers well clear of the certainties of ardour.

As the narrator warns in the "Prelude" to *Middlemarch*, "if there were one level of feminine incompetence as strict as the ability to count three and no more, the social lot of women might be treated with scientific certitude." Unfortunately, as the "Supreme Power" has endowed women's natures with an "inconvenient indefiniteness," that certitude eludes us (Eliot, *Middlemarch* 4). The novel that ensues makes the limitations faced by women clear, especially in the "Finale" when the narrator tells us in so many words that the "determining acts" of Dorothea's life were "the mixed result of a young and noble impulse struggling amidst the conditions of an imperfect social state" (838). That "imperfect social state" is evidently the one in operation in Eliot's own time and place. "A new Theresa will hardly have the opportunity of reforming a conventual life," the narrator explains, "any more than a new Antigone will spend her heroic piety in daring all for the sake of a brother's burial: the medium in which their ardent deeds took shape is for ever gone" (838). If Eliot admires a more exciting past, still the stark clarity of "ardent deeds" was never her aim. As Phyllis Rose has argued, despite her famous act of cohabitation, Eliot, with her insistence that friends refer to her as "Mrs. Lewes," was committed less to the radical restructuring of Victorian society than to her relationship with one individual, George Henry Lewes. In her own rendering, Eliot's chosen "medium," the nineteenth-century novel, also tells a story of trade-offs: what Dorothea's story loses in glamour, it gains in contingency, uncertainty, and second chances.

While *Middlemarch* is especially notable for the tidy pairs of women and men that apparently bring the novel to an end, the marriage of one woman and one man as both problem and answer is a marker of the nineteenth-century novel more generally, whether taken as the British "family" novel, the novel of adultery, or as the unequal power dynamic that so often drives the novel of "sensation." Of the novelists considered here, neither Tolstoy nor Hawthorne is reliably on the side of women's equality, Tolstoy as he holds to an ideal of women as wives and mothers alone, and Hawthorne with his creeping mockery of Zenobia and, behind her, the real Margaret Fuller. Even so, the fates of Tolstoy's various heroines, especially Dolly Oblonsky, can't but call into question the viability of traditional marriage as a means of structuring society, while Hawthorne is simply inconsistent. Whatever Hawthorne's reflections on Zenobia, *The Scarlet Letter*, published two years before *The Blithedale Romance* but set a full two centuries earlier, ends not just with Hester Prynne's re-incorporation into society on her own terms, but with her perhaps over-optimistic belief that "in some brighter period, when the world should have grown ripe for it, in Heaven's own time, a new truth would be revealed, in order to establish the whole relation between man and woman on a surer ground of mutual happiness" (Hawthorne, *Scarlet* 189). Although the mostly female novelists of sensation, like Eliot, were much more pointed in their criticism of the status quo, again of the authors discussed here, it is only Chernyshevsky in *What Is to Be Done?* who spells out in some detail what Hawthorne's "surer ground" might entail: although her eventual accomplishments as a trained doctor in no way match Eliot's real accomplishments as a writer, still Vera Pavlovna's carefully choreographed move from her first husband to her second offers a practical means of achieving the sexual freedom that we also glimpse in the discreet flitting from room to room in her fourth dream. Dostoevsky, like Eliot, stops short of easy answers. Even so, Nina Pelikan Straus sets him equally firmly in the feminist camp.

As Straus argues in *Dostoevsky and the Woman Question* (1994), despite Dostoevsky's advocacy of an orthodox and so patriarchal form of Christianity, it is no coincidence that Raskolnikov brutally murders two women, that the horse in the dream is a mare, that Svidrigailov commits crimes against women, or that Raskolnikov just happens to encounter a group of prostitutes, "almost every one of" whom "had a black eye" (*Crime* 157; *PSS* 6: 122). Straus also draws our attention to the role of women in the novel so pointedly titled *The Brothers Karamazov*, and especially to Katerina Ivanovna's and Grushenka's refusal to reconcile one with another in the novel's penultimate chapter. Even as in the very last chapter "[t]he all-boy audience cheers the arrival of the good father Alyosha, and women are decidedly invisible," Straus writes, still "Katya's indeterminate configuration in the text's Christological structure produces the central loophole in the narrative. She and Ivan are the 'individuals' in the novel whose independence presents the greatest challenge to Alyosha's (and Dostoevsky's) final solution" (Straus 140). Certainly, Dostoevsky has other ways of rendering the wrong relationship of self to other, in the mutually destructive relationship of prison guards and prisoners in *Notes from the House of the Dead* (*Записки из мертвого дома*, 1862), for example, or in the dialectic of masters and servants in *Notes from Underground* and *The Brothers Karamazov*. As already noted in chapter three, however, Dostoevsky most often demonstrates the murderous effects of hierarchical relationships in gendered terms.

In *Middlemarch*, Dorothea starts by assuming the same principle of inequality, not just in her idealization of Casaubon, but in the very notion that "[t]he really delightful marriage must be that where your husband was a sort of father" (Eliot, *Middlemarch* 10), while Lydgate with his fundamental belief in "the innate submissiveness of the goose as beautifully corresponding to the strength of the gander" (356) makes the same mistake twice, first with the French actress Laure, then with Rosamund. In Dostoevsky, it is the women who die, in Eliot, the men, sometimes in real terms, sometimes metaphorically. Either way, the fatal effects of inequality are the same. In Will, Dorothea finds a sibling, not a father, and if the reader still balks at her final domesticity even when cast as the realization of the names that she has borne all along, first and last, Eliot gives us a better or at least easier example of fellowship in the second couple joined in marriage at the end *Middlemarch*. As Rose argues, it is Fred Vincy and Mary Garth with the authorship of their books wrongly attributed each to the other who best approximate the very real fellowship that Marian Evans seems to have found with George Henry Lewes. For all her demonstrated ability to make her own way in the world, Eliot evidently needed a man, and she went out and got one, and the fruit of that relationship was not just her own writing, but also Lewes's, including the partly posthumously published *Problems of Life and Mind* that Eliot completed after his death. As Eliot's nod to Antigone and the conventual reforms of the real St. Theresa would suggest, however, fellowship also lends itself to relationships of a less heteronormative sort.

As I argue in a not-quite-conclusion that both recapitulates my argument and sets it in motion once more, for all the importance of marriage to the nineteenth-century novel, fellowship is not just an answer to the "Woman Question," but more broadly a way of being in the world with humans and non-humans alike. Even in *Middlemarch*, while Dorothea's third great act of giving achieves true sisterhood only very fleetingly, it is only that extraordinary moment of communion with Rosamund that makes her marriage with Will possible. Fellowship as configured in larger and less conventional terms is also more readily apparent in Collins and especially Dostoevsky. In *The Woman in White*, Fosco's "extraordinariness" is foiled not just by his "weakness" for Marian but by the still less conventional threesome of Walter, Laura, and Marian working, living, and loving together. In *The Brothers Karamazov*, while heteronormative couplings end on an inconclusive note, Ivan's love for the "sticky little leaves that come out in spring" (*Brothers* 230; *PSS* 14: 210) is realized instead in his love for his brothers: all of them.

Last Words and Open Endings

Epilogues and afterwords of all sorts operate in contradictory terms as an apparently final, more conclusive ending that nonetheless sits just beyond the original ending and so inevitably opens our reading back up. If a tension between open and closed is accordingly always in play, still the "Finale" to *Middlemarch* is particularly striking in its refusal of finality, as any expectations that the title itself might engender are promptly dashed by the very first sentence. As the "Finale" begins: "[e]very limit is a beginning as well as an ending" (*Middlemarch* 832). The narrator's most immediate point of reference is a non-ending of an extra-literary sort, as the "Finale" goes on to present the two marriages that unfold just beyond the novel proper as Eliot's ideal of ongoing, equal, and interactive fellowship. Marriage, as the narrator explains, and as the further adventures of the two happy couples would illustrate, although "the bourne of so many narratives," is also "the beginning of the home epic – the gradual conquest or irremediable loss of that complete union which makes the advancing years a climax, and age the harvest of sweet memories in common" (832). The narrator then waits until almost the very end to undermine her own emphasis on marriage with a sharp turn in a different direction.

To quote from the wistful not-quite-end once more, "[a] new Theresa will hardly have the opportunity of reforming a conventual life, any more than a new Antigone will spend her heroic piety in daring all for the sake of a brother's burial" (*Middlemarch* 838). These almost-last words send us all the way back to the "Prelude" where Saint Theresa is first introduced as a "little girl walking forth one morning hand-in-hand with her still smaller brother" (3).[1] The narrator's now framing of her novel, front and back, with reference to sisters real as well as chosen belatedly re-directs our attention away from Will and towards Dorothea's own intense experience of sisterhood. It is Celia, after all,

Dorothea's real, biological sister, who has been her companion all along and whose insistence on maintaining that relationship in the aftertime of the "Finale" helps provide for Dorothea's happy end. It is also a sister of the heart that Dorothea finds in Rosamund, if only for one brief, shining moment.

Deeply hurt by what she takes to be evidence of Rosamund's affair with Will, Dorothea nonetheless makes up her mind to return to the Lydgates' home the next day to support Rosamund in Lydgate's time of need. Eliot's steadily increasing distance from the religiosity of her youth is evident not just in the irregularity of her life and the early translation of *The Life of Jesus, Critically Examined*, but even at her death in her burial not in Westminster Abbey, but in Highgate Cemetery, alongside her common-law husband and not far from Karl Marx. Just as Dostoevsky never operates in Christian terms alone, however, so Christian imagery is not entirely absent from what Auerbach calls "the secular world of Middlemarch," as the fleeting moment of fellowship that ensues is framed with clear if quiet reference to the religion that Eliot apparently left behind. Not having slept all night, Dorothea opens her curtains at dawn:

> On the road there was a man with a bundle on his back and a woman carrying her baby; in the field she could see figures moving – perhaps the shepherd with this dog. Far off in the bending sky was the pearly light; and she felt the largeness of the world and the manifold wakings of men to labour and endurance. She was a part of that involuntary, palpitating life, and could neither look out on it from her luxurious shelter as a mere spectator, nor hide her eyes in selfish complaining (Eliot, *Middlemarch* 788).

As she readies herself to face Rosamund, her maid is struck by her face, "which in spite of bathing had the pale cheeks and pink eyelids of a *mater dolorosa*" (780). Dorothea carries that vision of the holy family, and of her own role as Mary, into her encounter with Rosamund.

Like many a Dostoevskyan character in need, Rosamund starts convinced of her own injury and armed in her own pride; as she understands it, "Dorothea was not only the 'preferred' woman, but had also a formidable advantage in being Lydgate's benefactor" (Eliot, *Middlemarch* 793). Under the influence of the freely expressed emotion that "had wrought itself more and more into … [Dorothea's] … utterance, till the tones might have gone to one's very marrow" (795), however, Rosamund experiences a change of heart. For a brief moment, "[p]ride was broken down between these two" (796), as Dorothea explains that what is hard about marriage is "the nearness that it brings." "Even if we loved some one else better than – than those we were married to, it would be no use," Dorothea says brokenly. "I mean, marriage drinks up all our power of giving or getting any blessedness in that sort of love" (797). As Rosamund is carried away by Dorothea's emotion, "the two women clasped each other as if they had been in a shipwreck" (797), and she tells Dorothea that what she thinks she saw wasn't so. It is only with this breaking down of the walls that separate her from Rosamund that Dorothea then finds her way to Will.

Even as instantiated in the institution of heterosexual marriage, this human interrelationship and even permeability of self and other may seem to operate on too small a scale to offer anything in the way of a real political program. When not cast as conventional marriage, in Eliot as in Dostoevsky, it may seem a little otherworldly. Just as science would seem to require a solution to all problems "once and for all," so politics, it might seem, is the French Revolution and the Paris Commune, Büchner and Vogt in Germany in 1848, Nechaev's *Catechism of a Revolutionary* (1869), Eliot with her eventual turn to Toryism, Dostoevsky's friendship with Ober-Procurator K.P. Pobedonostsev (1827–1907), chief advocate of reactionary politics under Alexander III. The problem is, as always, how we hold in our grasp as both idea and act the mutual implication of mind and body and mind and material world, abstract idea and concrete realization, tenor and vehicle, that the novels that have been my focus here, above all Dostoevsky's, both describe and embody.

I write these words in a time of great crisis, as Russian claims to "extraordinariness" have re-emerged in a particularly deadly combination of nationalist rhetoric and imperialist ambitions. As financed by our dependence on fossil fuels and with the threat of nuclear disaster ever-present, the full-scale invasion of Ukraine on 24 February 2022 also encapsulates our accelerating global climate crisis in especially urgent and terrifying form. Like far too many of the canonical figures in Russian literature, Dostoevsky with his undying faith in Russia's "great mission in the future renewal of mankind" readily lends himself to the project of Russian aggression. Even as we acknowledge the uses that Putin finds for him, however, Dostoevsky's own investment in this other strain of nineteenth-century science and literature, the non-positivist/nihilist sort, points in another direction.

Collins is another whose cohabitation with both Caroline Graves and Martha Rudd and, for the last twenty years of his life, with both at the same time, seems to have shaped his own unconventional views of social and political structures. Unlike Dickens, who, despite his own separation and subsequent affair, relegates Esther in *Bleak House* to the role of a deeply repressed Angel in the House, Collins gives his heroines considerably more agency. In *The Moonstone* we are told repeatedly that Rachel Verinder isn't like other girls, not just because she is the mistress of a considerable fortune but, more importantly, because she has the strength and determination to go her own way even in the face of significant social opprobrium. That we have a novel at all is only because Rachel with her unfeminine ways, as Betteredge writes, "never asked your advice; never told you beforehand what she was going to do; never came with secrets and confidences to anybody, from her mother downwards" (Collins, *Moonstone* 58). Were Rachel to have said what she saw when she saw it, we would have had not a novel in thirteen parts as presented by eleven different narrators, none of them Rachel, but a short story. In Franklin Blake, Rachel finds a husband perfectly ready to respect her independence, her secrecy, and her self-will, not that Collins can't imagine a more radical re-fashioning of patriarchy. In *The Moonstone* Limping Lucy in her bitter resentment at Franklin Blake's unwitting impact on the life of a poor servant girl dreams of a world without men altogether; of Rosanna Spearman she says, "I had a plan for our going to London together like sisters, and living by our needles" (191). In *The Woman in White*, Collins proposes not a couple, but a threesome.

Marian with the beautiful figure of a woman and, as Collins would have it, the ugly face, sharp intelligence and dauntless courage of a man, doesn't just cross lines of gender and sexuality, but together with Walter and Laura reimagines what family might entail. While it is Walter and Laura who fall in love and eventually marry, the original relationship, the one that sees Laura through her darkest days, is the unbreakable bond between the two (half-)sisters, Laura and Marian. Marian and Walter finally enjoy their own independent and intense connection as evident not just in their joint care for Laura at the novel's end, but in Marian's prophetic dream of Walter's return. *The Woman in White* would suggest that if two people can engage in a relationship of true equality, so can three. In *The Brothers Karamazov*, we expand from two to many.

Despite the example of Raskolnikov and Sonia in *Crime and Punishment*, when Dostoevsky imagines a dialogic relationship of equality, he most often does without heteronormative relationships altogether. In *Demons*, fellowship takes the form not just of a tragically short-lived holy family – Shatov, his wife Marya, and the baby that isn't his, all dead by the novel's end – but Shatov together with his sometime-comrade Kirillov, who, when not raving of the man-God, plays with the baby next door and who immediately turns to Shatov's aid when his pregnant wife unexpectedly returns. As Shatov says, "Kirillov! If … if you could renounce your terrible fantasies and drop your atheistic ravings … oh, what a man you'd be, Kirillov!" (*Demons* 571; *PSS* 10: 436). In *The Idiot*, for all the marriageability of the three Yepanchin sisters, not to mention Nastasya Filippovna, the central relationship is the connection between Myshkin and Rogozhin, light and dark, that doesn't hold. In *The Brothers Karamazov*, there's a woman for every brother, although no one marries and the role of women in the new world, as Straus points out, is left a little unclear. What we have instead are brothers who operate at all times on multiple levels at once, as the four biological (half-) brothers, the monks at the local monastery, the boys joined around Alyosha and the memory of Ilyusha at the novel's end, and the brotherhood of man.

Dostoevsky's last and longest novel might seem to offer the final word on fellowship, not just when the boys gather around Alyosha at the novel's very end, but in Alyosha's role throughout. Alyosha, although not a nihilist, and, indeed, while apparently lacking a political tendency altogether, is as ardent as any of Schiller's heroes. While the narrator is quick to assure us that his young hero "was not at all a fanatic, and, in my view at least, even not at all a mystic" (*Brothers* 18; *PSS* 14: 17), that he feels obliged to pre-empt any ideas that we might have otherwise is because Alyosha has been known to adopt extreme positions, not just his current intention to become a monk, but also in the "wild, frantic modesty and chastity" (20; *PSS* 14: 19) that made him the butt of a loving mockery while in school. As the narrator again explains at the start, and as the novel as a whole will make clear, what saves Alyosha from foolish or even misguided excess is his innate receptivity together with his fortunate meeting at a still young age with the Elder Zosima, "to whom he became attached with all the ardent [горячею] first love of his unquenchable heart" (18; *PSS* 14: 18). It is Zosima who directs Alyosha's Karamazovian "sensuousness" back out into the world where his biological brothers dwell and where he will one day marry Lise. It is also Zosima, in death as in life, who instils in Alyosha the love for a material world that in turn enjoys its own agency.

As an exhausted Alyosha grieves Zosima's death and bodily corruption, he dozes off to the sound of Father Paissy reading of the marriage at Cana, and his disconnected thoughts merge into a dream of the resurrected Zosima welcoming him to the very feast that Paissy's words describe. As Zosima in the dream invites Alyosha to look on the God, "[s]omething burned in Alyosha's heart, something suddenly filled him almost painfully, tears of rapture nearly burst from his soul" (*Brothers* 362; *PSS* 14: 327), he gives a "short cry," "w[a]ke[s] up," and steps outside to find himself one not just with his own mind and body but with world around him:

> Over him the heavenly dome, full of quiet, shining stars, hung boundlessly. From the zenith to the horizon the still-dim Milky Way stretched its double strand. Night, fresh and quiet, almost unstirring, enveloped the earth. The white towers and golden domes of the church gleamed in the sapphire sky. The luxuriant autumn flowers in the flowerbeds near the house had fallen asleep until morning. The silence of the earth seemed to merge with the silence of the heavens, the mystery of the earth touched the mystery of the stars. (362; *PSS* 14: 328)

As the night, the church, even the sleeping flowers reveal through their own "living lives" Beer's "clustering mystery of a material universe," Alyosha falls to the earth and embraces it as he "vow[s] ecstatically to love it, to love it unto ages of ages." In that moment, as the narrator explains with reference to Zosima's earlier teaching, "[i]t was as if threads from all those innumerable worlds of God all came together in his soul, and it was trembling all over, 'touching other worlds'" (362; *PSS* 14: 328). While Zosima's presence both in his own material body and as reflected in Alyosha's plays an essential part in eliciting this sense of fellowship with the world around him, still, as Feuerbach would have it, "[o]nly to an open mind does the world stand open" (Feuerbach, *Fiery* 241). It is Alyosha's good fortune that both his mind and his heart have been remarkably so all along.

Alyosha as we meet him in the novel offers a rare example of effective giving, as, while not right away, still he succeeds in giving to the impossibly proud Snegiryov, just as in his own time of need prior to returning to his father's home he "never worried about who was supporting him." "In this," the narrator explains, "he was the complete opposite of his older brother, Ivan Fyodorovich, who lived in poverty for his first two years at the university, supporting himself by his own labor, and who even as a child was bitterly aware that he was eating his benefactor's bread" (*Brothers* 21; *PSS* 14: 20). Alyosha instead befriends his fellow human beings at every turn, not just his brothers and Captain Snegiryov, but Grushenka, Lise, Kolia Krasotkin, and the boys at Kolia's school. As the boys shout at the novel's very end, "Hurrah for Karamazov!" and Dostoevsky evidently expects that his readers will, too. Even as he offers up Alyosha as a uniquely positive hero, his one example of a "positively good and beautiful man," however, Dostoevsky leaves the possibilities for fellowship open.

While the last scene points to an ideal of how we might be in and with the world, the indeterminacy that Dostoevsky set in motion in the novel's very first pages remains in effect, and different readers unfold that ending in different ways. For those who remember the second, "main" novel still to come, the one that Dostoevsky promised in his "Note from the Author," the last scene is not an ending at all, but, like Eliot's "Finale" or the Epilogue to *Crime and Punishment*, the start of a new story. For another, more Nabokovian reader, the jolly boys all joined around the living Alyosha and the dead Ilyushechka end on a note so cloying as to empty the conclusion of any meaning. A reader less invested in her own place in the canon of Russian literature might note that generic shifts are a distinct feature of the "irresolution" that has been Dostoevsky's aim throughout. For Nabokov himself, it is exactly this sort of scene that allowed him to dismiss Dostoevsky as "a much overrated, sentimental and Gothic novelist of the time" (Pushkin 3: 191). As Straus points out, this apparently ultimate moment of fellowship is finally also strikingly lacking in anyone not male. Difficult, even troubling as they may be, these "loop-holes" as they lend themselves to different readings reflect the "provisionality, openness, revisability, and collective endeavor" of that strain of nineteenth-century materialism that has been my starting point here.

As Greta Matzner-Gore argues, the claim to an open ending is commonplace in Dostoevsky scholarship post-Bakhtin, although one that is itself often left a little unexamined. As critics we tend to leave unclear whether the lack of closure that we have in mind pertains to the characters' fates, the readers' interpretative options, or both together; we often also simply take for granted the combined ethical and aesthetic value of "open-ness," not just for our own reading, but for Dostoevsky's writing, this when Dostoevsky's own stated intentions often call that value into question. As Matzner-Gore asks, "How can Bakhtin's position be reconciled with Dostoevsky's claim – recalled by a coworker – that his political foes 'do not suspect that the end of everything is coming soon ... to all their 'progress and chatter! They have no idea that the Antichrist has already been born ... and *is coming*.... And the end of the world is close, closer than they think'?" (Matzner-Gore, *Dostoevsky* 43).[2] A nineteenth-century science (small "s") as a set of shifting signs that bear an existential relationship with what they represent, as accordingly itself both open and closed, offers a way that we might answer that question. Lest I depart from that model and cast my own literary science a little too narrowly, however, not just for my reader, but for the ever-evolving reality that we share, I give my own last word not to Alyosha, but to his future mother-in-law, Mme. Khokhlakov, the wealthy Russian owner of an estate just outside Kharkiv.

Although Mme. Khokhlakov is a figure of fun from start to finish, it's not that her unfailingly silly statements never indicate anything true. As a drunken Razumikhin says in *Crime and Punishment*, "I like it when people lie! Lying is man's only privilege over all other organisms. If you lie – you get to the truth!" (*Crime* 202; *PSS* 6: 156). While Mme. Khokhlakov's lies are almost always a performance of self, she often also misses the point, most notably in her encounter with Dmitri just before his father's murder. Mme. Khokhlakov seems to be only half-listening when she frames her response to Dmitri's desperate plea for financial assistance in terms of her own disappointment that Zosima's dead body actually stinks. "[H]ere there's no question of presentiments, no retrograde pretense to miracles ... this, this is mathematics" (*Brothers* 384; *PSS* 14: 347), she tells Dmitri before making the absurd, entirely abstract, and yet apparently premeditated proposal that he set off without delay in search of gold mines. As she explains in what seems to be an attempt at abduction, "I even studied your gait and decided: this man will find many mines" (384; *PSS* 14: 348). With a forced smile, Dmitri asks, "From my gait?" "And why not from your gait," she says:

> What, do you deny that it's possible to tell a man's character from his gait, Dmitri Fyodorovich? Natural science confirms it. Oh, I'm a realist now, Dmitri Fyodorovich. From this day on, after all that story in the monastery, which upset me so, I'm a complete realist, and want to throw myself into practical activity. I am cured. Enough! as Turgenev said. (384–5; *PSS* 14: 348)

Mme. Khokhlakov's truth lies not in gold mines nor in the absurdity of her science, Peircean or otherwise, nor even in Dostoevsky's joke now at Turgenev's expense, but in the impossibility of engaging in any "practical activity" when we start with that paradoxical creation of the "vulgar" mind, a fully rational world made up of matter alone.

Like the many nineteenth-century writers and scientists whose still cutting-edge thought shaped his own, Dostoevsky argues that we do better to set minds in bodies and the two together as part of a material world that is both the object of our consideration and a subject in its own right. To do so is to recognize that science (small "s") gives us answers that we can and should use as we work to address the various degrees of material apocalypse that confront us, from the COVID pandemic to the war in Ukraine to climate change, just answers that won't serve "once and for all." Students of the climate crisis have their own term for this scientific as well as political starting point, not "fellow-ship" or "dialogicity," but "kinship," as in Kyle Whyte's contribution to *The Cambridge Companion to Environmental Humanities* (2021), "Time as Kinship." Whyte's is a complicated project, one rendered still less attainable by a Dostoevskyan commitment to Russianness as a kind of universality. As Dostoevsky himself reminds us, however, what we do in the world is not a given. In the real world as in its fictional render-ings, the signs that ground our actions shift: what and how we read matters.

Notes

Introduction: Dostoevsky, Science, and the Nineteenth-Century Novel

1 A particularly Russian reading is most often noted with regard to Darwin, see Daniel Todes, *Darwin without Malthus: The Struggle for Existence in Russian Evolutionary Thought* (Oxford University Press, 1989) and, as inspired by Todes, Liza Knapp's "Darwin's Plots, Malthus's Mighty Feast, Lammennais's Motherless Fledglings, and Dostoevsky's Lost Sheep," in *Dostoevsky beyond Dostoevsky* (Academic Studies Press, 2016).

2 Wolfe's term, taken from Humberto Maturana's influential work, is not "writing," but "autopoiesis."

3 On art and aesthetics as an "exemplary form of human meaning-making" (Johnson 212), see, for example, Mark Johnson, *The Meaning of the Body: Aesthetics of Human Understanding* (2007) and note that Johnson in turn points to John Dewey in *Art as Experience* (1934). As for claims for the nineteenth-century novel in particular, see, for example, Kay Young's claim in *Imagining Minds: The Neuro-Aesthetics of Austen, Eliot and Hardy* (2010): "the novel is an aesthetic map to and experience of the nature of the mind-brain" (K. Young 9).

1. Nineteenth-Century Materialisms

1 Chernyshevsky took his undergraduate degree in 1850 and later aspired to a master's degree; the higher degree failed to materialize, however, when his 1855 thesis, "The Aesthetic Relations of Art to Reality," was not accepted. In the meantime the death of his parents forced Dobroliubov to drop out of the university after only one year of study.

2 Piderit, not Piederit as Pevear has it. Piderit was trained as a doctor and physiologist and wrote on mimicry; his article here is "Gehirn und Geist," translated into Russian as "Мозг и его деятельность." Adolf Wagner, as

Pevear notes, was an economist inspired by Quetelet to turn his attention to "moral statistics."

3 While George Henry Lewes argued vigorously for the other side (and also engaged in the dissection of frogs himself), British anti-vivisectionists, as Jessica Straley writes, "counted on their side a veritable 'who's who' of Victorian authors, among them John Ruskin, Matthew Arnold, Alfred Tennyson, Robert Browning, Lewis Carroll, Christina Rossetti, Vernon Lee, and Ouida" (Straley 354–5).

4 Virtanen quotes Bernard's contemporary Louis Ménard, actually a one-time chemist and also a writer and artist: "Claude Bernard's widow has opened a home for dogs to make amends for the crimes of experimental physiology. On Judgment Day, this offering will weigh more in the awesome balance than all the discoveries of her husband" (Virtanen 119).

5 The catalogue of the Russian National Library lists two early lithographic editions of *Kraft und Stoff* (in Russian, *Сила и вещество*), one published in Moscow in 1860 by the "Moscow student circle of Zaichnevskii and Agripopulo," and another listed as lacking both title page and cover and given an admittedly approximate but also impossible date of "1850 (?)."

6 Cherno can only speculate that Feuerbach was joking in an article devoted not so much to Moleschott, nor even to his own materialist philosophy, but to a belated interest in politics, hence the title of the piece, "Science and Revolution" ("Die Naturwissenschaft und die Revolution," 1850).

7 Because Feuerbach emphasizes the role of the imagination (see below), Aleksei Vdovin argues that Chernyshevsky's debt to Feuerbach is overrated, see his "Chernyshevskii vs. Feierbakh: (psevdo) istochniki dissertatsii 'Esteticheskie otnosheniia iskusstva k deistvitel'nosti." Like Aileen Kelly, I would make the broader case that we have been misreading Feuerbach altogether.

2. Of Doctors and Detectives, Chemistry and Mathematics

1 For more on Eliot's double time frame, see Gillian Beer, "What's Not in Middlemarch."

2 Louise McReynolds offers many examples of the increasing importance of forensic science in the Russian context, including, in a case that Dostoevsky famously refers to in *The Idiot*, in the 1867 trial of the student Danilov where "a cold-blooded killer was identified in part by a cut on his left hand" (L. McReynolds 54).

3 Note that, where Rakitin in Dmitri's account invites science in in the person of Claude Bernard, Raskolnikov's particular touchstone is Napoleon, as in Zamyotov's blurted-out question: "Might it not have been some future Napoleon who bumped off our Alyona Ivanovna with an axe

last week?" (*Crime* 266; *PSS* 6: 204), or in Raskolnikov's own later musing: "No, these people are made differently; the true *master*, to whom all is permitted, sacks Toulon, makes a slaughter house of Paris, *forgets* an army in Egypt, *expends* half a million men in a Moscow campaign, and gets off with a pun in Vilno; and when he dies they set up monuments to him – and thus *everything* is permitted" (274; *PSS* 6: 211). For more on Napoleon in *Crime and Punishment*, see Lindenmeyr.

4 While it's apparently Ivan who first uttered the words, we hear them first in an account given by Dmitri's distant cousin and sometime benefactor Pyotr Aleksandrovich Miusov. According to Miusov, Ivan at a recent local gathering informed his fellow guests that love is a matter only of our belief in our immortality, and that were that belief to be destroyed, "not only love but also any living power to continue the life of the world would at once dry up in it. Not only that, but then nothing would be immoral any longer, everything would be permitted, even anthropophagy" (*Brothers* 69; *PSS* 14: 64–5). Note also that, while Ivan (like Raskolnikov) then confirms Miusov's version of what he said, it is still without explicit reference to "permission": "Yes, it was my contention," he says, "There is no virtue if there is no immortality" (*Brothers* 70; *PSS* 14: 65).

5 In my highly condensed argument here I draw on Winter, Taylor, and Pearl but come to slightly different conclusions.

6 Both well-known, but nonetheless very different. As Jenny Bourne Taylor notes, "[i]n conflating Carpenter and Elliotson in this way Collins is condensing two figures whose names … would have had very different resonances in the 1860s: Carpenter, the respected voice of mainstream physiological psychology; Elliotson, the marginalized advocate of mesmerism" (Taylor 183).

7 As for detectives, clues, and ratiocination at its crudest, see S.S. Van Dine's "Twenty Rules for Writing Detective Stories" (1928).

8 In his journal *Time* in 1861 Dostoevsky published three stories by Poe in Russian translation, "The Tell-Tale Heart" (1843), "The Black Cat" (1843), and "The Devil in the Belfry" (1839), together with a brief preface.

9 For a very thorough accounting of calculus in *War and Peace*, see Love.

10 Named after German mathematician Bernhard Riemann (1826–66), also a major player in the development of non-Euclidean geometry.

11 As for hedgehogs, note also the example of Casaubon with his *Key to All Mythologies* in *Middlemarch*.

3. Bodies and Plots

1 See, for example, when Ivan's devil responds to the question of his own reality with mocking reference to the love of plot that Tolstoy would deny.

"In dreams and especially in nightmares," the devil says, "well, let's say as a result of indigestion or whatever, a man sometimes sees such artistic dreams, such complex and real actuality, such events, or even a whole world of events, woven into such a plot, with such unexpected details, beginning from your highest manifestations down to the last shirt button, as I swear even Leo Tolstoy couldn't invent" (*Brothers* 639; *PSS* 15: 74).

2 Note that B.G. Reizov associates the proposed elopement in *The Village of Stepanchikov* not with the novel of "sensation" but with Dickens; for a detailed comparison of this scene with the episode in *The Pickwick Papers* when Mr. Jingle attempts to run off with Miss Rachel Wardle, see his "K voprosu o vliianii Dikkensa na Dostoevskogo."

3 See, for example, "'Sasha, you didn't even let me get to what I wanted to talk about,' Vera Pavlovna began some two hours later when they sat down to tea" (Chernyshevsky, *What* 344).

4 Bernard's blithe way with vivisection was learned at the feet of his mentor and medical school professor, leading French physiologist François Magendie. As Jerome Tarshis explains, one day Magendie brought a live duck to class. As his lecture came to an end, Tarshis writes, Magendie turned to the duck: "'We shall take advantage of the few minutes we have left,' he said, 'to remove the cerebral lobes of a duck…. I am removing the top of the duck's skull by a stroke of the scalpel,' he continued. 'Then I scrape away the lobes of the brain. You see that more skill than time is required. Now I draw the skin together again to protect what is left in the skull from injury or foreign bodies. It is extraordinary how little harm this experiment does to these birds. A duck with its head practically empty does not differ, in gait and behavior, from a duck with its brain intact'" (Tarshis 26–7).

5 If Chernyshevsky glosses over the fatal implications of his science of life, note that many of his most articulate readers were quick to see the cost that his vision of future happiness entails. In his 1864 review of the novel, for example, Chernyshevsky's sometime ally M.E. Saltykov-Shchedrin joked that "'with time' the dear nihilists will cut through human corpses with an impassive hand and at the same time take up dancing and singing" (Drozd 2). Strakhov, on the other hand, saw nothing funny in an underlying principle of death. As he wrote in an article intended for Dostoevsky's journal *Epoch*, "I don't need your kind of happiness! For this inhuman coldness, for this terrible emptiness which you call happiness, for this infallibility and indestructible tranquility I would not give up at any price my present life which is hard and meager but nevertheless full and warm and pulsing" (Drozd 16–17).

6 Even Latimer's curious blindness to the thoughts of one woman, Menke notes, is echoed in Lydgate's choice of Rosamund.

7 Beer notes the echo of Huxley in these lines, see Beer, *Darwin's Plots* 142. Note also the Bernardian challenge that they contain. As Flint asks, "if we could have more strongly developed powers of vision, if we could lay bare the future and the thoughts of others, as a physiologist can lay bare the hidden workings of the mysteriously veiled human body, would we choose to accept such powers?" ("Blood" 457).

8 As Matzner-Gore argues, readers since have continued to criticize Dostoevsky on ethical grounds; as she summarizes Dmitri Merezhkovsky's response, for example, "Dostoevsky's 'insatiable curiosity' about the 'most frightening and shameful ulcers of the human soul' crosses a moral and aesthetic line" (Matzner-Gore, *Dostoevsky* 16).

9 Note that Dostoevsky's God exactly fills what Shuttleworth describes as the "logical gap" in Lewes's and Eliot's thought.

10 Although Sechenov seems not to have been particularly interested in effecting social change, his naturally occurring nihilist credentials are impeccable. While Sechenov himself expressed nothing in the way of political aims, in the wake of Karakazov's attempt to assassinate Alexander II in 1866, the censorship committee in St. Petersburg attempted judicial procedures, accusing Sechenov in *Reflexes of the Brain* of not just of spreading materialism but of the "debasing" of Christian morality. His personal life is still more striking. As Irina Paperno explains, Sechenov's not yet life partner, Mariia Obrucheva, first entered into a fictitious marriage with Pyotr Bokov and then fell in love with her liberator, only to end by falling in love with her physiology professor instead – who then settled down with the Bokovs in the early 1860s in a friendly *ménage à trois*. Interestingly, while Chernyshevsky seems to have known of Obrucheva and Bokov, the rest of the story unfolded while he was already in prison and before the publication of his novel; as Paperno writes, it is "a remarkable case of the mutual influence of life and literature" (Paperno 135).

11 In "Dmitry Grigorovich and the Limits of Empiricism" Matzner-Gore claims that Dostoevsky "[f]rom the very beginning of his career … used Grigorovich's work as the foil against which to define himself as an artist" (Matzner-Gore, "Empiricism" 375–6); as she notes, Grigorovich and Dostoevsky met while still in school and then shared an apartment as struggling young writers in 1844–5. Matzner-Gore nonetheless argues that Dostoevsky's criticism just misses the mark, as Grigorovich in her reading is less interested in objective truth than in the limits of what we can know.

12 "Cold and dead materialism" is Hawthorne's description of Westervelt's mesmerism in *Blithedale Romance*, see my chapter two.

4. Metaphor and Allegory

1 Coleridge is following up on Goethe's famous distinction between allegory "wo das Besondere nur als Beispiel, als Exempel des Allgemeinen gilt" ["where the particular serves only as an example of the general"], and the truly poetic device of symbol, "wo das Besondere das Allgemeinere repräsentiert, nicht als Traum und Schatten, sondern als lebendig-augenblickliche Offenbarung des Unerforschlichen" ["where the particular represents the more general, not as a dream or a shadow, but as a living momentary revelation of the Inscrutable"] (Fletcher 13, fn 24).

2 In the original Russian: "Трудно было бы представить более жалую, более пошлую, более бездраную и пресную аллегорию" (*PSS* 10: 389). Note that the governor's wife herself describes her "самым экономическим, немецким балком," as "единственно для аллегории" ["economical little German ball ... solely as an allegory] (*PSS* 10: 356; *Demons*, 465).

3 While Menand is here describing another member of the Metaphysical Club, Peirce's friend Nicholas St. John Green, the way of thought is shared. As Menand summarizes: "The 'proximate cause' is just the antecedent event people choose to pick out in order to serve whatever interest they happen to have in the case at hand" (Menand 223).

4 Lest we think that we know the one way to interpret these already complicated images, the same contrast abruptly softens in *The Brothers Karamazov* when Dmitri in his final scene proposes that he escape to America only to return once he's mastered English "как самые что ни на есть англичане" (*PSS* 15: 186) ["as well as any downright Englishman" (*Brothers* 765)]; in Dostoevsky's last and longest novel, it is Lopukhov again, but America is no longer an image of death alone.

5 As Tresch explains, geographer and naturalist Alexander Humboldt (1769–1859) relied on a dizzying array of measuring devices understood not as "a transparent means of registering nature 'in itself,'" but as extensions of himself or even as animate objects in their own right. "Instruments that made good traveling companions – those that were small, light and versatile – were favored, like the portable barometer that could be fitted onto the head of Humboldt's walking stick" (Tresch 79), Tresch writes, while "his most cherished compasses, barometers, and sextants were discussed with the same enthusiastic affection as his dearest friends" (78). For more on Humboldt's "republic of instruments," see Tresch on Schiller in chapter five.

6 As Ian Duncan notes parenthetically, "(Shepherds would be unlikely in Middlemarch, an agricultural and commercial district)" (Duncan 150).

It says something for the "Russian point of view" that Duncan himself sees in this reference only a "radical naturalization" at work, not an admittedly subtle nod to the story of Christ.

7 As A.N. Kaul writes, "Blithedale itself, as we shall see, is finally judged in terms of its own professed values and not by the standards and norms of society. It is only when, and insofar as, the visionaries themselves turn out to be men of iron masquerading in Arcadian costume, that Blithedale is dismissed as humbug.... But the individualism he champions is not incompatible with, but rather tends toward and finds its richest fulfillment in, the human community" (Hawthorne, *Blithedale* 307).

8 On more meanings of Coverdale's name, see Britt, "The Veil of Allegory in Hawthorne's *The Blithedale Romance*."

9 Kate Holland has similarly argued that Lebedev's laughable claims in *The Idiot* in no way undermine the significance of his apocalyptic vision, and I am still responding to her "Hurrying, Clanging, Banging and Speeding for the Happiness of Mankind: Railways, Metaphor and Modernity in *The Idiot*" (presentation, Annual Convention of the American Association for Slavic, East European and Eurasian Studies, Chicago, IL, 9–12 November 2017).

5. Social Physics

1 While Count Fosco plays a double game, his ultimately right-wing tendencies are evident in what would seem to be his betrayal of the revolutionary group that also includes Walter's friend Pesca; Holmes, on the other hand, presents a more complicated case in his politics as in his science. We know that Conan Doyle himself was an advocate of the British Empire, and the thread of empire that runs all through Holmes's cases – treasure from India, both clients and criminals from America and Australia, even Watson with his injury sustained in Afghanistan – would seem to associate Holmes's "superior powers of observation" and "use of the scientific method" with the imperial project. That said, Holmes opens himself also to other readings; see, for example, O'Dell's "Performing the Imperial Abject: The Ethics of Cocaine in Arthur Conan Doyle's *The Sign of Four*" (2012).

2 According to McLean, the 1830s were an "era of widespread admiration for the Poles"; it was only by mid-century that "radical groups were co-opting the Polish cause, producing images of international solidarity that perhaps appealed to working-class and radical readers but frightened everyone else." See McLean 164, 160.

3 On Dorothea and Margaret Fuller, see Deery.

4 For a transcript of Putin's remarks at the annual Valdai meeting in October 2022, including his reference to Dostoevsky, see http://kremlin.ru /events/president/news/69695 (accessed 19 August 2024). Note that the theme of the 2022 meeting was "A Post-Hegemonic World: Justice and Security for Everyone."

5 From an 1880 letter to a friend: "At ten I saw in Moscow a performance of 'The Robbers' with Mochalov acting one of the principal roles. I can justly say that the strong impression of this performance has acted as an enormous stimulation for my entire spiritual development" (Kostka 215). The show opened in 1829 and made a regular part of Mochalov's repertoire into the 1830s.

6 While Schiller's theme of parricide serves as an obvious plot device in *The Brothers Karamazov*, Alexandra Lyngstad identifies Schilleresque motifs throughout all Dostoevsky's novels, ranging from the "ecstasy of spite" to the *"Brüderschaft-murder complex"* and even the image of the piano key in *Notes from Underground*; Susan McReynolds traces the influence of Schiller's aesthetics in Dostoevsky's 1861 essay "Mr. -bov and the Question of Art."

7 As for Herzen's final, again mixed, relationship with Schiller, I quote from Kostka: "The one hundredth anniversary of Schiller's birth, on November 10, 1859, was commemorated by celebrations throughout Europe and the rest of the civilized world. Commenting on the preparations for such a celebration in England, Herzen writes in a letter to the historian Michelet, under the date of November 4, 1859: On se prepare (même ici) [à] fêter le jour de naissance de *Schiller* – dans le Crystal Palace. C'est bien que les grands hommes de l'Allemagne sont morts – et qu'ils ne peuvent assister eux-mêmes à leur jubilés." As Kostka adds, "[i]n spite of his disgust, Herzen prevailed upon himself to attend the celebration. A letter to his son reflects his disappointment with the whole affair" (Kostka 175).

8 Schiller's first dissertation was rejected on the score that it was a little too freewheeling and even "fiery," and the following year he submitted two more on two very different topics, "On the Difference between Inflammatory and Putrid Fevers" ("De differentia febrium inflammatorium et putidarum") and "Essay on the Connection between the Animal and the Spiritual Nature of Man," probably, as Dewhurst and Reeves argue, in the hopes that at least the former would pass muster. As it turned out, while Schiller's faculty required that Schiller cut "several imaginative and flowery passages (50), they passed the essay on animal and spiritual nature; the less controversial work on fevers was again rejected.

9 As Malia notes, the quasi-exception is William Tell.

Conclusion: Last Words and Open Endings

1 Scholars of Eliot are quick to connect this image of the two siblings hand-in-hand with Maggie Tulliver's death by drowning together with her brother Tom in *The Mill on the Floss* (1860) as well with the real Marian Evans's relationship with her older brother Isaac; while the two were close as children, Isaac cut off all contact during the years of his sister's cohabitation with Lewes.
2 The co-worker whom Matzner-Gore mentions is once more V.V. Timofeeva, the young and politically progressive proofreader that Dostoevsky worked with as editor of *The Citizen* (*Гражданин*) in 1873–4, see Dolinin 170.

Works Cited

Abram, David. *The Spell of the Sensuous: Perception and Language in a More-Than-Human World*. New York: Vintage Books, 1997.

Ahearn, Stephen T. "Tolstoy's Integration Metaphor from War and Peace." *The American Mathematical Monthly* 112, no. 7 (2005): 631–38.

Auerbach, Nina. "Dorothea's Lost Dog." In *Middlemarch in the Twenty-First Century*, edited by Karen Chase, 87–106. Oxford: Oxford University Press, 2006.

Avdeeva, L. R. *Russkie Mysliteli: Ap. A. Grigor'ev, N. Ia. Danilevskii, N.N. Strakhov: Filosofskaia kul'turologiia vtoroi poloviny xix veka*. Moskva: Izd-vo Moskovskogo universiteta, 1992.

Bakhtin, M. M. *The Dialogic Imagination: Four Essays*. Edited by Michael Holquist. Translated by Caryl Emerson. Austin: University of Texas Press, 1981.

Balzac, Honoré de. *Père Goriot*. Translated by A. J. Krailsheimer. Oxford: Oxford University Press, 1991.

Barthes, Roland. *Camera Lucida: Reflections on Photography*. Translated by Richard Howard. New York: Hill and Wang, 1981.

Barzun, Jacques and Wendell Hertig Taylor. *A Catalogue of Crime*. New York: Harper & Row, 1971.

Beer, Gillian. *Darwin's Plots: Evolutionary Narrative in Darwin, George Eliot and Nineteenth-Century Fiction*. Cambridge: Cambridge University Press, 2009.

– *George Eliot*. Bloomington, IN: Indiana University Press, 1986.

– "What's Not in Middlemarch." In *Middlemarch in the Twenty-First Century*, edited by Karen Chase, 15–36. Oxford: Oxford University Press, 2006.

Belknap, Robert. *The Genesis of The Brothers Karamazov: The Aesthetics, Ideology, and Psychology of Making a Text*. Evanston, IL: Northwestern University Press, 1990.

– *The Structure of the Brothers Karamazov*. The Hague: Mouton, 1967.

Bennett, Jane. *The Enchantment of Modern Life: Attachments, Crossings, and Ethics*. Princeton, NJ: Princeton University Press, 2001.

– *Vibrant Matter: A Political Ecology of Things*. Durham: Duke University Press, 2010.

Berlin, Isaiah. *The Hedgehog and the Fox: An Essay on Tolstoy's View of History*. New York: Simon and Schuster, 1970.

Bernard, Claude. *An Introduction to the Study of Experimental Medicine*. Translated by Henry Copley Greene. New York: Dover Publications, 1957.

Bethea, David and Victoria Thorstensson. "Darwin, Dostoevsky, and Russia's Radical Youth." In *Dostoevsky Beyond Dostoevsky: Science, Religion, Philosophy*, edited by Svetlana Evdokimova and Vladimir Golstein, 35–62. Brighton, MA: Academic Studies Press, 2016.

Bishop, Paul. "'Elementary Aesthetics', Hedonist Ethics: The Philosophical Foundations of Feuerbach's Late Works." *History of European Ideas* 34, no. 3 (2008): 298–309.

Blix, Göran. "The Occult Roots of Realism: Balzac, Mesmer, and Second Sight." *Studies in Eighteenth-Century Culture* 36 (2007): 261–80.

Britt, Brian M. "The Veil of Allegory in Hawthorne's *The Blithedale Romance*." *Literature and Theology* 10, no. 1 (1996): 44–57.

Brodhead, Richard. "Hawthorne and the Fate of Politics." *Essays in Literature* 11, no. 1 (1984): 95–103.

Brooks, Peter. *Reading for The Plot*. Cambridge, MA: Harvard University Press, 1992.

Brunson, Molly. *Russian Realisms*. DeKalb: Northern Illinois University Press, 2016.

Buckle, Henry Thomas. *History of Civilization in England*. 2 vols. New York: D. Appleton and Co., 1925.

Campbell, Lewis. *The Life of James Clerk Maxwell*. New York: Johnson Reprint Corp., 1969.

Cherno, Melvin. "Feuerbach's 'Man Is What He Eats': A Rectification." *Journal of the History of Ideas* 24, no. 3 (1963): 397–406. https://doi.org/10.2307/2708215.

Chernyshevskii, N. G. *Estetika*. Moscow: Khudozhestvennaia literatura, 1958.

– *Selected Philosophical Essays*. Moscow: Foreign Languages Publishing House, 1953.

Chernyshevsky, Nikolay Gavrilovich. *What Is to Be Done?* Translated by Michael R. Katz. Ithaca: Cornell University Press, 1989.

Collins, Wilkie. *The Moonstone*. New York: Modern Library, 2001.

– *The Woman in White*. London: Oxford University Press, 1975.

Comte, Auguste. *Auguste Comte and Positivism: The Essential Writings*. Edited by Gertrude Lenzer. New Brunswick, NJ: Transaction, 1998.

Daly, Nicholas. *Literature, Technology, and Modernity, 1860–2000*. Cambridge: Cambridge University Press, 2004.

Davis, Michael. *George Eliot and Nineteenth-Century Psychology: Exploring the Unmapped Country*. Aldershot: Ashgate, 2006.

Deery, Patricia. "Margaret Fuller and Dorothea Brooke." *The Review of English Studies, New Series* 36, no. 143 (August 1985): 379–85.

Dewhurst, Kenneth and Nigel Reeves. *Friedrich Schiller, Medicine, Psychology and Literature: With the First English Edition of His Complete Medical and Psychological Writings*. Berkeley: University of California Press, 1978.

Dickens, Charles. *Bleak House*. New York: Modern Library, 2002.

Dolinin, A., ed. *Dostoevskii v vospominaniakh sovremennikov*. Vol. 2. Leningrad: Khudozhestvennaia literatura, 1964.

Dostoevskii, F. M. *Polnoe sobranie sochinenii v tridtsati tomakh*. Leningrad: Nauka, 1972–90.

Dostoyevsky, Fyodor. *The Adolescent*. Translated by Richard Pevear and Larissa Volokhonsky. New York: Knopf, 2003.

– *The Brothers Karamazov*. Translated by Richard Pevear and Larissa Volokhonsky. New York: Farrar, Straus and Giroux, 2002.

– *Crime and Punishment*. Translated by Richard Pevear and Larissa Volokhonsky. New York: Alfred A. Knopf, 1993.

– *Demons*. Translated by Richard Pevear and Larissa Volokhonsky. New York: Alfred A. Knopf, 1995.

– *The Idiot*. Translated by Richard Pevear and Larissa Volokhonsky. New York: Vintage, 2003.

– *Notes from Underground*. Translated by Richard Pevear and Larissa Volokhonsky. New York: Alfred A. Knopf, 1993.

– *The Village of Stepanchikovo and Its Inhabitants*. Translated by Ignat Avsay. Ithaca: Cornell University Press, 1987.

Doyle, Arthur Conan. *The Complete Sherlock Holmes*. London: CRW Publishing, 2005.

Drozd, Andrew Michael. *'Chernyshevskii's What Is to Be Done?: A Reevaluation*. Evanston, IL: Northwestern University Press, 2001.

Druzhinin, A. V. "Novosti angliiskoi literatury." In *Sobranie sochinenii*. Vol. 5. St. Petersburg: Imperatorskaia akademii nauk, 1865.

Duncan, Ian. *Human Forms: The Novel in the Age of Evolution*. Princeton: Princeton University Press, 2019.

Eco, Umberto. *The Limits of Interpretation*. Bloomington: Indiana University Press, 1990.

– *Postscript to The Name of the Rose*. San Diego: Harcourt Brace Jovanovich, 1984.

Eliot, George. *Daniel Deronda*. Oxford: Oxford University Press, 1984.

– *The Essays of George Eliot*. Edited by Thomas Pinney. New York: Columbia University Press, 1963.

– *The George Eliot Letters.* Edited by Gordon S. Haight. 9 vols. New Haven: Yale University Press, 1954–78.

– *Middlemarch.* London: Penguin, 1994.

Engels, Friedrich. "Old Preface to Dühring. On Dialectics." In *Karl Marx and Friedrich Engels: Collected Works.* Vol. 25. New York: International Publishers, 1975.

Fantina, Richard. *Victorian Sensational Fiction: The Daring Work of Charles Reade.* New York: Palgrave Macmillan, 2010.

Feuerbach, Ludwig. *The Essence of Christianity.* Translated by George Eliot. Amherst, NY: Prometheus Books, 1989.

– *The Fiery Brook; Selected Writing of Ludwig Feuerbach.* Translated with an Introd. by Zawar Hanfi. Garden City, NY: Anchor Books, 1972.

"Fiziologiia obydennoi zhizni. Soch. G. G. L'iusia." *Vremia* 6 (1861): 50–63 (p. 51).

Fletcher, Angus. *Allegory: The Theory of a Symbolic Mode.* Ithaca, NY: Cornell University Press, 1964.

Flint, Kate. "Blood, Bodies, and the Lifted Veil." *Nineteenth-Century Literature* 51, no. 4 (March 1997): 455–73.

– "The Materiality of Middlemarch." In *Middlemarch in the Twenty-First Century*, edited by Karen Chase, 65–86. Oxford: Oxford University Press, 2006.

Frank, Joseph. *Dostoevsky: The Mantle of the Prophet, 1871–1881.* Princeton, NJ: Princeton University Press, 2002.

– *Dostoevsky: The Seeds of Revolt, 1821–1849.* Princeton, NJ: Princeton University Press, 1976.

Frank, Lawrence. *Victorian Detective Fiction and the Nature of Evidence.* Houndmills, Basingstoke, Hampshire: Palgrave Macmillan, 2009.

Gaffney, Peter, ed. *The Force of the Virtual: Deleuze, Science, and Philosophy.* Minneapolis: University of Minnesota Press, 2010.

Goldstein, Amanda Jo. *Sweet Science: Romantic Materialism and the New Logics of Life.* Chicago: University of Chicago Press, 2017.

Gregory, Frederick. *Scientific Materialism in Nineteenth Century Germany.* Dordrecht, Holland: D. Reidel, 1977.

Grossman, Leonid. *Balzac and Dostoevsky.* Ann Arbor, MI: Ardis, 1973.

Gubailovskii, Vladimir. "Geometriia Dostoevskogo: Tezisy k issledovaniiu." In *Roman F. M. Dostoevskogo 'Brat'ia Karamazovy'*, edited by T. A. Kasatkina, 39-69. Moscow: Nauka, 2007.

Guth, Deborah. *George Eliot and Schiller: Intertextuality and Cross-Cultural Discourse.* Hampshire: Ashgate, 2003.

Haight, Gordon. *George Eliot: A Biography.* New York: Oxford University Press, 1968.

Harrowitz, Nancy. "The Body of the Detective Model: Charles S. Peirce and Edgar Allan Poe." In *The Sign of Three: Dupin, Holmes, Peirce*, edited by Umberto Eco and Thomas A. Sebeok, 179–97. Bloomington: Indiana University Press, 1983.

Hawthorne, Nathaniel. *The Blithedale Romance*. New York: W. W. Norton, 1978.

– *The Complete Novels and Selected Tales of Nathaniel Hawthorne*. Edited by Norman Holmes Pearson. New York: Modern Library, 1937.

– *The Scarlet Letter*. New York: W. W. Norton, 1962.

Helmholtz, Hermann von. *On the Sensations of Tone as a Physiological Basis for the Theory of Music*. New York: Dover, 1954.

Henry, Nancy. "George Eliot and Politics." In *The Cambridge Companion to George Eliot*, edited by George Levine, 138–58. Cambridge: Cambridge University Press, 2001.

Hofstadter, Douglas R. *Gödel, Escher, Bach: An Eternal Golden Braid*. New York: Basic Books, 1979.

Holquist, Michael. "Whodunit and Other Questions: Metaphysical Detective Stories in Postwar Fiction." In *The Poetics of Murder: Detective Fiction and Literary Theory*, edited by Glenn W. Most and William W. Stowe, 149–74. San Diego: Harcourt Brace Jovanovich, 1983.

Horváth, Géza S. "The Great Sinner, the Harlot and the Eternal Book: The Genre of Romance as Plot and Textual Motif in Dostoevsky's *Crime and Punishment* in Comparison with Hawthorne's *Scarlet Letter*." *The Dostoevsky Journal* 20 (2019): 31–53.

Hutter, Albert D. "Dreams, Transformations, and Literature: The Implications of Detective Fiction." *Victorian Studies* 19, no. 2 (December 1975): 181–209.

Jakobson, Roman. "On Realism in Art." In *Language in Literature*, edited by Roman Jakobson and Stephen Rudy. Cambridge, MA: Belknap Press, 1987.

James, Henry. "Current Literature." *The Galaxy* 15, no. 3 (March 1873): 422–29.

James, William. *Varieties of Religious Experience*. Cambridge, MA: Harvard University Press, 1985.

Johnson, Barbara. "The Frame of Reference: Poe, Lacan, Derrida." *Yale French Studies* 55–56 (1977): 457–505.

Johnson, Mark. *The Meaning of the Body: Aesthetics of Human Understanding*. Chicago: University of Chicago Press, 2007.

Kaladiouk, A. S. "On 'Sticking to the Fact' and 'Understanding Nothing': Dostoevskii and the Scientific Method." *The Russian Review* 65 (3): 417–38.

Kasatkina, Tat'iana. "Lazarus Resurrected: A Proposed Exegetical Reading of Dostoevsky's *Crime and Punishment*." *Russian Studies in Literature: A Journal of Translations* 40, no. 4 (Fall 2004): 6–37.

Katz, Michael. "Dostoevsky and Natural Science." *Dostoevsky Studies* 9 (1988): 63–77.

Kazakov, A. A. "F. M. Dostoevskii i L. Feierbakh: Ob odnom iz vozmozhnykh istochnikov dialogizma Dostoevskogo." *Vestnik Tomskogo gosudarstvennogo universiteta* 361 (2012): 13–16.

Kelley, Theresa. *Reinventing Allegory*. Cambridge: Cambridge University Press, 1997.

Kelly, Aileen. *The Discovery of Chance: The Life and Thought of Alexander Herzen*. Cambridge, MA: Harvard University Press, 2016.

Kiiko, E. I. "Vospriiatie Dostoevskim neevklidovoi geometrii." In *Dostoevskii: materialy i issledovaniia*. Vol. 6, 120–28. Leningrad: Nauka, 1985.

Kitzinger, Chloë. *Mimetic Lives: Tolstoy, Dostoevsky, and Character in the Novel*. Evanston, IL: Northwestern University Press, 2021.

Knapp, Liza. *The Annihilation of Inertia: Dostoevsky and Metaphysics*. Evanston, IL: Northwestern University Press, 1996.

– "Darwin's Plots, Malthus's Mighty Feast, Lammennais's Motherless Fledgelings, and Dostoevsky's Lost Sheep." In *Dostoevsky Beyond Dostoevsky*, edited by Sveltana Evdokimova and Vladimier Golstein, 63–82. Brighton, MA: Academic Studies Press, 2016.

– "Realism." In *Dostoevsky in Context*, edited by Deborah A. Martinsen and O. E. Maiorova, 229–35. Cambridge: Cambridge University Press, 2016.

Kostalevsky, Marina. "Sensual Mind: The Pain and Pleasure of Thinking." In *A New Word on "The Brothers Karamazov"*, edited by Robert Jackson, 200–09. Evanston, IL: Northwestern University Press, 2004.

Kostka, Edmund K. *Schiller in Russian Literature*. Philadelphia, PA: University of Pennsylvania Press, 2016.

Krieger, Murray. "'A Waking Dream': The Symbolic Alternative to Allegory." In *Allegory, Myth, and Symbol*, edited by Morton W. Bloomfield, 1–22. Cambridge, MA: Harvard University Press, 1981.

Kuleshov, V. I., ed. *Fiziologiia Peterburga*. Moscow: Nauka, 1991.

Lakoff, George and Mark Johnson. *Metaphors We Live By*. Chicago: University of Chicago Press, 1980.

Lampert, E. *Sons against Fathers: Studies in Russian Radicalism and Revolution*. Oxford: Clarendon Press, 1965.

Latour, Bruno. *Pandora's Hope: Essays on the Reality of Science Studies*. Cambridge, MA: Harvard University Press, 1999.

Levine, George. *Dying to Know: Scientific Epistemology and Narrative in Victorian England*. Chicago: University of Chicago Press, 2002.

Lewes, George Henry. *The Physiology of Common Life*. 2 vols. New York: D. Appleton, 1860; repr. 1969.

– *Problems of Life and Mind: First Series, the Foundations of a Creed*. 2 vols. Boston: Houghton, Osgood, 1875–80.

– *Ranthorpe*. Athens: Ohio University Press, 1974.

– "What Is Sensation?" *Mind* 1, no. 2 (1876): 157–61.

Lindenmeyr, Adele. "Raskolnikov's City and the Napoleonic Plan." *Slavic Review* 35 (1976): 37–47.

Lotman, Jurij and Boris Uspenskij. *The Semiotics of Russian Culture*. Edited by Ann Shukman. Ann Arbor: University of Michigan Press, 1984.

Love, Jeff. "Tolstoy's Calculus of History." *Tolstoy Studies Journal* 13 (2001): 23-37.

Lyngstad, Alexandra H. *Dostoevsky and Schiller*. Mouton de Gruyter: Mouton, Humanities, 1975.

Malia, Martin E. *Alexander Herzen and the Birth of Russian Socialism, 1812–1855*. Cambridge, MA: Harvard University Press, 1961.

Malmgren, Carl Darryl. *Anatomy of Murder: Mystery, Detective, and Crime Fiction*. Bowling Green, OH: Bowling Green State University Popular Press, ©2001.

Mansel, H. L. "Sensation Novels." *Quarterly Review* 113, no. 226 (April 1863): 482–514.

Maturana, Humberto and Francisco Varela. *The Tree of Knowledge: The Biological Roots of Human Understanding*. Boston: Shambhala, 1998.

Matzner-Gore, Greta. "Dmitry Grigorovich and the Limits of Empiricism." *The Russian Review* 77, no. 3 (2018): 359–77. https://doi.org/10.1111/russ.1218.

– *Dostoevsky and the Ethics of Narrative Form*. Evanston, IL: Northwestern University Press, 2020.

Maunder, Andrew, ed. *Varieties of Women's Sensation Fiction, 1855–1890*. Vol. 1. London: Pickering and Chatto, 2004.

Maxwell, James Clerk and W. D. Niven. *The Scientific Papers of James Clerk Maxwell*. Mineola, NY: Dover Publications, 2003.

McLean, Thomas. *The Other East and Nineteenth-Century British Literature: Imagining Poland and the Russian Empire*. New York: Palgrave MacMillan, 2012.

McReynolds, Louise. *Murder Most Russian: True Crime and Punishment in Late Imperial Russia*. Ithaca: Cornell University Press, 2013.

McReynolds, Susan. "Schillerian Aesthetic Humanism in *Vremia*." *Dostoevsky Studies* 7 (2003): 81–88.

Menand, Louis. *The Metaphysical Club*. New York: Farrar, Straus, and Giroux, 2001.

Menke, Richard. "Fiction as Vivisection: G. H. Lewes and George Eliot." *ELH* 67, no. 2 (Summer 2000): 617–53.

Merleau-Ponty, Maurice. *Phenomenology of Perception*. Translated by Colin Smith. New York: Humanities Press, 1962.

Mikhailovskii, N. K. "Zhestokii talant." In *F.M. Dostoevskii v russkoi kritike*, edited by A. A. Belkin, 306–85. Moscow: Khudozhestvennaia literatura, 1956.

Miller, D. A. *The Novel and the Police*. Berkeley: University of California Press, 1988.

Miller, Edwin Haviland. *Salem Is My Dwelling Place: A Life of Nathaniel Hawthorne*. Iowa City: University of Iowa Press, 1991.

Miller, J. Hillis. "The Two Allegories." In *Allegory, Myth, and Symbol*, edited by Morton W. Bloomfield, 355–70. Cambridge, MA: Harvard University Press, 1981.

Miller, Robin Feuer. "Adventures in Time and Space: Dostoevsky, William James, and the Perilous Journey to Conversion." In *William James in Russian Culture*, edited by Joan Delaney Grossman and Ruth Rischin, 33–58. Lanham, MD: Lexington Books, 2003.

Mochulsky, Konstantin. *Dostoevsky: His Life and Work*. Princeton, NJ: Princeton University Press, 1967.

Morson, Gary Saul. "Paradoxical Dostoevsky." *The Slavic and East European Journal* 43, no. 3 (1999): 471–94.

Nabokov, Vladimir. *The Gift*. Translated by Michael Scammell. New York: Vintage, 1991.

O'Dell, Benjamin D. "Performing the Imperial Abject: The Ethics of Cocaine in Arthur Conan Doyle's the Sign of Four." *The Journal of Popular Culture* 45, no. 5 (2012): 979–99. https://doi.org/10.1111/j.1540-5931.2012.00969.x.

Otis, Laura. *Membranes: Metaphors of Invasion in Nineteenth-century Literature, Science and Politics*. Baltimore: Johns Hopkins University Press, 1999.

Ovsianiko-Kulikovskii, D. N. "Dostoevskii v 70-kh godakh." In *Sobranie sochinenii*. Vol. 8, 204–18. St. Petersburg: Prometei, 1911.

Page, Norman, ed. *Wilkie Collins: The Critical Heritage*. London: Routledge, 1974.

Paperno, Irina. *Chernyshevsky and the Age of Realism: A Study in the Semiotics of Behavior*. Stanford, CA: Stanford University Press, 1988.

Paulsen, Kris. "The Index and the Interface." *Representations* 122, no. 1 (2013): 83–109.

Pearl, Sharrona. "Dazed and Abused: Gender and Mesmerism and Wilkie Collins." In *Victorian Literary Mesmerism*, edited by Martin Willis and Catherine Wynne, 163–82. Amsterdam: Rodopi, 2006.

Peirce, Charles S. *The Collected Papers of Charles Sanders Peirce 3/4*. Cambridge, MA: Harvard University Press, 1960.

– *The Essential Peirce: Selected Philosophical Writings*. Edited by Peirce Edition Project. 2 vols. Bloomington, IN: Indiana University Press, 1992–98.

– *Philosophical Writings of Peirce*. Edited by Justus Buchler. New York: Dover Publications, 1955.

Pisarev, D. I. *Polnoe sobranie sochinenii i pisem v 12-i tomakh*. Moscow: Nauka, 2001.

Poe, Edgar Allen. *The Complete Stories*. New York: Knopf, 1992.

Ponge, Francis. *The Power of Language: Text and Translations*. Translated by Serge Gavronsky. Berkeley: University of California Press, 1979.

Pushkin, A. S. *Eugene Onegin*. Translated by V. V. Nabokov. 4 vols. New York: Bollingen Foundation, 1964.

Reed, T. J. *Schiller*. New York: Oxford University Press, 1991.

Reitblat, A. I. "Detektivnaia literatura i russkii chitatel'." In *Knizhnoe delo v Rossii vo vtoroi polovine XIX-nachale XX vv*, 126–32. St. Petersburg: RNB, 1994.

Reizov, B. G. "К вопросу о влиянии Диккенса на Достоевского." *Язык и литература* 5 (1830): 153–70.

Rose, Phyllis. *Parallel Lives: Five Victorian Marriages*. New York: Knopf, 1983.

Ryan, Vanessa L. *Thinking Without Thinking in the Victorian Novel*. Baltimore: Johns Hopkins University Press, 2012.

Rylance, Rick. *Victorian Psychology and British Culture, 1850–1880*. Oxford: Oxford University Press, 2002.

Schiller, Friedrich. *Don Carlos*. Translated by Hilary Collier Sy-Quia and Peter Oswald. In *Don Carlos and Mary Stuart*, 1–202. Oxford: Oxford University Press, 2008.

– *Plays*. Edited by Walter Hinderer. New York: Continuum, 2003.

– *The Robbers*. Translated by F. J. Lamport. In *Sturm Und Drang*, edited by Alan C. Leidner, 181–298. New York: Continuum, 1992.

Sebeok, Thomas and Jean Umiker-Sebeok. "'You Know My Method': A Juxtaposition of Charles S. Peirce and Sherlock Holmes." In *The Sign of Three: Dupin, Holmes, Peirce*, edited by Umberto Eco and Thomas A. Sebeok, 11–54. Bloomington: Indiana University Press, 1983.

Sechenov, I. M. *Refleksy golovnogo mozga*. Moskva: Akademiia nauk, 1961.

Sharpe, Lesley. *Friedrich Schiller: Drama, Thought, and Politics*. Cambridge: Cambridge University Press, 1991.

Shkurinov, P. S. *Pozitivizm v Rossii XIX veka*. Moscow: Izd-vo Moskovskogo universiteta, 1980.

Shneyder, Vadim. "Kapitalizm kak povestvovatel'naia problema, ili Pochemu den'gi ne goriat v romane F. M. Dostoevskogo 'Idiot'." In *Russkii realizm XIX veka*, edited by M. Vaisman, A. Vdovin, I. Kliger and K. Ospovat, 272–95. Moscow: NLO, 2020.

Shuttleworth, Sally. *George Eliot and 19th Century Science: The Make-Believe of a Beginning*. New York: Cambridge University Press, 1984.

Stigler, Stephen M. *The History of Statistics: The Measurement of Uncertainty Before 1900*. Cambridge, MA: Harvard University Press, 1986.

Stone, Donald D. *The Romantic Impulse in Victorian Fiction*. Cambridge, MA: Harvard University Press, 1980.

Strakhov, N. N. *Mir kak tseloe*. Moskva: Airis Press, 2007.

Straley, Jessica. "Love and Vivisection: Wilkie Collins's Experiment in Heart and Science." *Nineteenth-Century Literature* 65, no. 3 (2010): 348–73.

Straus, Nina Pelikan. *Dostoevsky and the Woman Question*. New York: Palgrave Macmillan, 1994.

Tambling, Jeremy. *Allegory*. New York: Routledge, 2009.

Tarshis, Jerome. *Claude Bernard: Father of Experimental Medicine*. New York: Dial Press, 1968.

Taylor, Jenny Bourne. *In the Secret Theatre of Home*. London: Routledge, 1988.

Tegan, Mary Beth. "Strange Sympathies: George Eliot and the Literary Science of Sensation." *Women's Writing* 20, no. 2 (May 2013): 168–85.

Thomas, Ronald R. "*The Moonstone*, Detective Fiction and Forensic Science." In *Cambridge Companion to Wilkie Collins*, edited by Jenny Bourne Taylor, 65–78. Cambridge: Cambridge University Press, 2006.

Thompson, Diana Oenning. "Dostoevskii and Science." In *The Cambridge Companion to Dostoevskii*, edited by William J. Leatherbarrow, 191–211. Cambridge: Cambridge University Press, 2002.

Tkachev, P. N. "'Bol'nye liudi: 'Besy,' roman Fedora Dostoevskogo, v trekh chastiakh." In *Kritika 70-kh godov XIX veka*, edited by S. F. Dmitrenko, 67–122. Moscow: Olimp, 2002.

Tolstoi, L. N. *Polnoe sobranie sochinenii*. 90 vols. Moscow: Khudozhestvennaia literatura, 1935–64.

Tolstoy, Ivan. *James Clerk Maxwell: A Biography*. Chicago: University of Chicago Press, 1982.

Tolstoy, Leo. *War and Peace*. Translated by Richard Pevear and Larissa Volokhonsky. New York: Alfred A. Knopf, 2007.

Tresch, John. *The Romantic Machine: Utopian Science and Technology After Napoleon*. Chicago: University of Chicago Press, 2012.

Turgenev, Ivan. *Fathers and Children*. Translated by Michael Katz. New York: W.W. Norton, 2009.

Valentino, Russell. *Vicissitudes of Genre in the Russian Novel*. New York: Peter Lang, 2001.

Vdovin, Aleksei. "Chernyshevskii vs. Feierbakh: (psevdo) istochniki dissertatsii 'Esteticheskie otnosheniia iskusstva k deistvitel'nosti." *Zeitschrift für Slavische Philologie*, vol. 68, no. 1 (2011): 39–66.

– "Dostoevsky, Sechenov, and the Reflexes of the Brain: Towards a Stylistic Genealogy of *Notes from Underground*." In *Dostoevsky at 200*, edited Katherine Bowers and Kate Holland, 99–117. Toronto: University of Toronto Press, 2021.

Virtanen, Reino. *Claude Bernard and His Place in the History of Ideas*. Lincoln, NE: University of Nebraska Press, 1960.

Wellek, René. *Concepts of Criticism*. New Haven: Yale University Press, 1963.

Whittaker, Robert T. *Russia's Last Romantic, Apollon Grigor'ev, 1822–1864*. Lewiston: Edwin Mellen Press, 1999.

Winter, Alison. *Mesmerized: Powers of Mind in Victorian Britain*. Chicago: University of Chicago Press, 1998.

Wolfe, Cary. "Ecological Poetics." In *The New Wallace Stevens Studies*, edited by Bart Eeckhout and Gül Bilge Han. New York: Cambridge University Press, 2021.

Woolf, Paul. "Prostitutes, Paris, and Poe: The Sexual Economy of Edgar Allan Poe's 'the Murders in the Rue Morgue'." *Clues: A Journal of Detection* 25, no. 1 (2006): 6–19.

Woolf, Virginia. *The Common Reader*. New York: Harcourt, Brace and Co., 1925.

Young, Kay. *Imagining Minds: The Neuro-Aesthetics of Austen, Eliot and Hardy*. Columbus, OH: The Ohio State University Press, 2010.

Young, Sarah J. "Deferred Senses and Distanced Spaces: Embodying the Boundaries of Dostoevsky's Realism." In *Dostoevsky at 200*, edited by Katherine Bowers and Kate Holland, 118–36. Toronto: University of Toronto Press, 2021.

Zagidullina, M. V. "Dostoevskii glazami sootechestvennikov." In *Roman F. M. Dostoevskogo 'Idiot': Sovremennoe sostoianie izucheniia*, edited by T. A. Kasatkina, 508–39. Moscow: Nasledie, 2001.

Zola, Émile. *The Experimental Novel, and Other Essays*. Translated by Belle M. Sherman. New York: Cassell, 1893.

– *Thérèse Raquin*. Translated by Willard Trask. New York: Bantam Books, 1960.

Index